Understanding Animal Behaviour

Published by
Whittles Publishing,
Dunbeath,
Caithness KW6 6EG,
Scotland, UK

www.whittlespublishing.com

ISBN 978-184995-330-6

Printed by Short Run Press

Understanding Animal Behaviour

Rory Putman

Whittles Publishing

Contents

Preface

It is a problem, it would seem, with any rapidly advancing science that successive professional publications concern themselves increasingly with progressive refinement of the most modern, or recent, of theories and somehow lose the context in which these are rooted. Thus each generation of academic papers concerns itself only with the developments of theory of the previous two or three years – and tends increasingly to debate their niceties in isolation and out of the broader context of more general theory, which may remain largely unaffected. However, each new publication becomes so concerned with dotting 'i's and crossing 't's that, in paying attention to these minutiae, the main text is sometimes forgotten or, at least, often erroneously, considered as general knowledge by those too close to the finer detail.

As scientific publications become increasingly focused and erudite, more and more involved in what might appear to be a semantic debate of the subtlest detail, so these publications become inaccessible to the amateur or casual reader who merely seeks a simple overview of more general theory. Even such general texts as may be available in support of university courses often get bogged down in overmuch detail or complexity for the non-specialist reader.

The greatest risk in attempting an overview for that non-specialist, however, and in going back to basics, is that the text will be regarded as over-simplistic, naive or simply out-dated, by those specialists who do then pick it up (or worse, review it!). For a number of years I have been tempted to stick my neck out and compose such a book; for an equal number of years I have shied away from the task, waiting for someone more qualified and more articulate than myself to undertake the task. Yet maybe they, also, are too close to the fine detail – or too nervous of the risk of critical review by professional colleagues!

What has finally prompted me to cast caution to the wind – and put onto paper this basic introduction to animal behaviour for the interested non-specialist, was the recent sad task of editing my late father's papers. Amongst them was a series of philosophical essays in which he had tried to make sense of some 80 years' experience of life, as both man and scientist, and offer to himself some conclusions about life, the universe and everything. The essays were illuminating – but in many places sadly flawed by incomplete knowledge or understanding. My father was a successful scientist (a nuclear physicist of some reputation) and a lifelong naturalist – with a passionate interest in behaviour. Many of the essays indeed dealt with an analysis of human behaviour patterns by comparison with the behaviour of other animals, or at least interpreted within such a context; yet it was in these essays, to my surprise, that the logic was often marred by misconception or, at least, incomplete understanding of things which I myself would have assumed to be common knowledge – especially to one such as my father who, albeit as an amateur, had particular interest in behaviour and who read widely. If even his knowledge was so patchy and incomplete, I had to argue, then there must indeed be many others who would be glad of a simple book reviewing the very basics of our understanding of how and why animals behave as they do.

The current book, then, is a development from a series of essays or lectures of my own, tried and tested in numerous adult education courses given over the years to other intelligent and interested non-specialists like my father. It is not an academic textbook – indeed would be, and doubtless will be, scorned by many professionals as perhaps dated and certainly as over-simplistic. It uses words and concepts which are intuitively easy to grasp, even if they are in places oversimplified, in an attempt to offer to an amateur audience an insight into the way animals behave, but couched within a framework and in a conceptual language with which they can actually identify. I do not seek to patronise, but to inform.

If, in my attempts to make it more accessible, I irritate professional behaviourists by adopting devices or simplifications now frowned upon, or outmoded, I can only apologise – but to an extent accept such criticism in advance, since this book is aimed not at them but at that more general readership.

The book itself is presented in two main parts (although in practice these two sections are effectively independent, and may indeed be read in either order). In the first I seek to explore the 'how' of animal behaviour, taking a mechanistic approach to explore what we know of the way in which animals perceive their environment and what determines how they respond to it in the way that they do and what that response will be. In the latter part of the book, the focus switches more to the 'why' of animal behaviour, asking not so much how given responses are mediated through nerve and muscle systems, or how stimuli are perceived in the brain, but rather what has moulded why a particular behaviour takes the form that it does in its expression, what evolutionary forces have shaped – and continue to shape – the detailed form of more complex behaviours. Why do animals forage in the way they do? How may that foraging pattern be refined to optimal efficiency? Why do animals adopt the particular reproductive strategy and breeding behaviour that they do, (and why is there is such a bewildering variety of different ways of solving apparently the same problem)? Why do some animals live as solitary individuals, while others live in groups? The list goes on.

It might be expected that I would end with a chapter comparing animal and human behaviours, but this seemed far too simple, and probably far too contentious, even if it would have echoed my father's earlier attempts. Rather I would leave the reader to make their own analyses of the behaviour of animals around them, human and otherwise, and end instead with a series of questions and answers which have been asked of me over the years by those same 'interested amateurs'.

But this volume is hopefully more than simply a non-technical book about animal behaviour as such; or at least that is perhaps only part of the story. For the intention is that it should also be highly visual – and copiously illustrated throughout, in support and interpretation of the text. To an extent then, while the pictures are there to enhance understanding of the written text, the text itself also takes on an additional role as a showcase for some of the exquisite illustrations of my wife, wildlife artist Catherine Putman, who has agreed to collaborate with me on the project. The book is intended, to a degree, as a joint presentation of text and pictures, in celebration of the delights of a greater understanding of animals and the way they behave.

Acknowledgements

I would like to extend my thanks to Catherine Putman for her patience in preparing the numerous illustrations which accompany the text, as well as redrawing some of the diagrammatic figures.

Many of the illustrations in the early chapters (pages 18, 20, 22, 23 and 24) are based on the work of Professor Niko Tinbergen and his colleagues, and are redrawn from figures in his 1951 book *The Study of Instinct*; I am most grateful to his publishers, Oxford University Press, for permission to reproduce or redraw them here. Professor Geoff Parker and his publishers, John Wiley & Sons, kindly let me reproduce figures on the optimality of mating times in yellow dungflies (pages 80 and 81) from publications in 1974[1] and 1978[2]. Professor Henry Horn kindly gave permission for reproduction of his illustration of changing social organisation among Brewer's blackbirds in response to resource availability and distribution (page 90).[3] Illustration of the communication dances of honeybees on pages 61, 62 and 65 are redrawn with permission of Nature America, Inc. from Karl von Frisch's article: Dialects in the Language of the Bees (*Scientific American*, August 1962) and the diagram on page 99 of the way in which human hunter-gatherer societies respond to abundance and predictability of resources is redrawn from Dyson-Hudson and Smith (1978)[4] with permission of the authors and the American Anthropological Association. Full credits are listed in the captions of those figures. Antonio Uzal and Frauke Ohl helped me by redrawing the underlying graphs presented on pages 18, 29, 35 and 59. All other illustrations are original or re-drawn from the original sources; where I have been unable to establish the owner of copyright of those original sources I can only apologise that they are not fully credited here.

Rory Putman

1 Parker G.A. (1974) The reproductive behavior and the nature of sexual selection in *Scatophaga stercoraria*. *Evolution* 28: 93–108

2 Parker, G.A. (1978) Searching for mates. Ch. 8 in Krebs, J.R.and Davies N.B. (eds) *Behavioural Ecology*; reproduced in the 1981 edition of Krebs and Davies. *An Introduction to Behavioural Ecology*, Blackwell Scientific Publications.

3 Horn, H.S. (1968)The adaptive significance of colonial nesting in the Brewer's blackbird, *Euphagus cyanocephalus*. *Ecology* 49:682–694

4 Dyson-Hudson, R. and Smith, E. (1978) Human territoriality: an ecological re-assessment. *American Anthropologist* 80: 21–41.

1

Understanding animal behaviour

As animals ourselves, and whether we live in city or country, we live our lives surrounded by animals, both wild and domestic. Television – and the incredible skills of the modern wildlife photographer – beam a succession of the most amazing wildlife spectaculars from around the world, directly into our living rooms. And whether or not they may be keen naturalists, or have no specialist interest at all, most people have at one time or another observed animals around them, whether on film or in their own back garden, doing something unexpected – and may have wondered exactly how or why the creature acted as it did. This book is a brief attempt to explain something of what is going on. Its aim is to inform, to enhance awareness and understanding, but not to rob the natural world of its wonder. Be prepared, that it offers a formal, analytical approach to the study of behaviour – essential if we are properly to understand why, how and what animals are doing (without subjective anthropomorphic leaps of 'intuition'). But don't let that put you off: for if my own experience is anything to go by, formal study of the science merely enhanced my enjoyment, increased my wonderment.

In studying the behaviour of animals around us, we can, in practice, ask questions at two basic levels. Firstly, we may ask *why* an animal performs a given behaviour in a given set of circumstances, looking for the function of the behaviour in survival terms, why it evolved; why the behaviour actually takes the form that it does, rather than any of a host of alternatives; why some animals live in groups while others are solitary, what decisions animals must make in order to forage efficiently in different situations, and so on. But a different question asks *how* an animal performs a certain behaviour in a given set of circumstances: the physiological nuts and bolts, if you like, of how, mechanically, the response happens: how a given stimulus can cause, physiologically, a given response. And in this book we will adopt both approaches in trying to understand both the how and also the why.

The first part of the book will concern itself with 'how': how particular actions are put together, how they are mediated through nerve and muscle systems and how they are co-ordinated: how it is, in effect, that an animal responds to a given situation with appropriate behaviour. This rather mechanistic approach aims to unravel the basics of how animals behave and this first chapter presents a brief overview of the whole, to try and establish an overall framework. We will revisit these same ideas in greater detail in later chapters, but at least we may start with an initial outline of the whole, to which that later detail may be attached.

Defining the parameters of study

What exactly is behaviour? Formally, behaviour can be defined as an animal's perception of, and response to, its environment. By this, though, we do not mean solely its physical or ecological context; in addition to its physical surroundings, the animal's environment includes, of necessity, other living organisms around it – of its own and of different species: potential mates or potential competitors, predators and prey.

The response the animal gives to some facet of its surroundings can be relatively simple (perhaps a simple orientation of the body with respect to some environmental cue) or more complex. Necessarily, the response involves some form of 'motor response' (expressed in muscle activity, or perhaps more subtly in the release of specific hormones). Very few behaviours involve only a single motor action, so the response usually also involves the co-ordination of a number of separate motor actions, simultaneously or sequentially, in what is referred to as a full behavioural response or 'action pattern'. This means that even the simplest behaviour patterns we may observe are in fact composed of a number of functional units linked together and co-ordinated into a single pattern.

Yet such complete behaviour patterns are by definition too complex to be easily analysed and understood with our current mechanistic approach. We will in the second part of the book attempt to understand *why* these behaviours take the form they do, in terms of the function they perform, but in our current analysis of how the behaviours are controlled and put together, there are too many things involved, too many things going on. In order to study the behaviour, to try to understand what is going on, it is necessary to break such complex patterns down into their component building blocks and try and understand their mechanics, before building them back up again into the full response (and in that rebuilding, trying to understand how the separate building blocks are coordinated and integrated).

The separate motor action patterns of a behavioural response (muscle responses and, occasionally, hormonal changes) are more the realm of physiology than behaviour in that, reduced to extremes, we should be considering in our analyses the way in which electrical changes across cell membranes result in the transmission of messages down nerve pathways, and how similar changes in the electrical potential of the membranes of muscle fibres result in their contraction. I do not intend, however, to get into the finer detail here of such neuro-muscular physiology. For us, the realm of behaviour 'accepts' that nerve and muscles systems work (by whatever means) and that animals indeed do, on occasions, secrete hormones into their bloodstream that may have a more general influence on their internal state.

In these pages I am more concerned with

i) how those responses are triggered in the first place (i.e. how they are released by the animal's perceptions of its environment) and
ii) how they may be modified and linked together: the different levels of co-ordination of these basic muscle spasms into more complicated behaviour.

If we can understand these two things, then we'll have a pretty good idea of what's involved and should be able to interpret and explain the occurrence of almost any behaviour pattern you may observe. And, for me, the acid test of understanding is the ability not only to explain what has been observed, but also to predict what an animal will do next. Accurate prediction, after all, is a true test of whether you really understand what is going on. And in essence: it's very simple.

The building blocks of behaviour

At the very simplest level of behaviour, certain behaviour patterns may be identified as pure behavioural reflexes. These are, in essence, very similar in a behavioural sense to purely **physiological reflexes**. Given some specific cue from the environment, the animal will respond, predictably (and automatically) with a specific response; repeatedly, and always with the same response, like simple automata.

These reflexes can be linked together to form chains of actions, with each action acting as the releasing stimulus for the next action and so on. They are still rigid and inflexible (because each action triggers the next in the sequence in a fixed relationship) but nonetheless they can result in apparently quite complex patterns of behaviour (lots of linked motor actions). Reflexes are therefore not restricted to a simple one-stimulus:single motor action response. However, because each action acts as the cue for the next in the series, these chains of behaviour are inflexible, unchangeable and predetermined. They are carried through without any internal control every time the correct releasing stimulus is observed. Indeed, they are commonly referred to as **Fixed Action Patterns**.

Lest we should dismiss these behaviours as unsophisticated and perhaps suitable for controlling the behaviour of only the simplest of organisms, it is worth noting that the very inflexibility of such reflex chains, together with the fact that they do not have to be 'thought about' and are thus performed extremely rapidly in response to the appropriate cues, makes them ideal for controlling simple survival behaviours, such as escape from danger. Indeed, this type of response has huge advantages in a number of instances: it's pre-programmed, fast-acting and immediate. Survival behaviours, after all, do not necessarily need fancy refinement: if, as an animal, you are threatened by a predator, it is indeed better not to have to stop and think about escaping, but just get out of there…

At this level of individual actions or fixed action patterns, as long as we know what any animal's repertoire of fixed action patterns is and what the reflex links are (i.e. which stimuli promote which response), and as long as we know the sensory capacity of our animal (i.e. what stimuli it can actually perceive) then we can place an animal in any given environment and predict exactly what it will do. However, such a system lacks flexibility in terms of whether or not to respond at all (there is no element of choice: it's a hard-wired reflex) and also of how to respond. In production of more complex behaviours and responses, such reflexes can be modified in various ways.

Modification of response

First, for example, we may consider how those fixed patterns may be modified by **learning**. By experience, in learning, new associations of stimulus and response may be put together, so that the animal learns in a given situation that it is appropriate to deliver a given response. Alternatively, old associations may be altered: an old response previously performed in response to some given stimulus may be linked instead to a new stimulus (as in the classic and well-known example of Pavlov's dogs on page 52).

The effects of learning, however, still do not prevent us from analysing the mechanics of the observed behaviour pattern, nor from predicting what an animal's behaviour will be in a given situation. As long as we know what experiences our animal has had – and what associations it is likely to have learnt – we can still model its likely future behaviour quite accurately. And for most relatively simple animals which lack any great individuality, the environment in which they develop and live is likely to be much the same for all individuals in any given species, so they are all likely to learn much the same sort of things. We can therefore still predict behaviour, without knowing very much about our *individual* animal's past history.

Simple behaviours can be modified in other ways too. Performance or non-performance of particular behaviours can be modified, for example, by consideration of an animal's internal state. If a domestic dog has recently drunk, it will not need to drink again for a while and can therefore switch off, as it were, the action pattern 'water-drink' so that, for a while, it is unresponsive to the stimulus of water. Expression of a whole host of other behaviours can also be modified and controlled in the form of their expression by the animal's current internal state.

Such phenomena can be included in our 'nuts and bolts' mechanistic model of behaviour by allowing some consideration of the animal's **motivational state**. Its internal motivation to perform or not to perform a given action may thus modify or completely suppress the simple expected stimulus-response reflex. But even so the animal's internal state is itself inevitably a consequence or a function of its own immediate past behaviour, (since behaviour by definition is the only way it interacts with its environment and can change its internal state). So if we can make some guess at its internal state from a study of its ongoing behaviour, we can predict how this internal state will affect its responsiveness – and can thus still predict its future behaviour, if not with absolute certainty, then still with a high degree of accuracy. We still understand why it reacts in the way it does.

We are already a long way from simple reflexes, but we can still explain what is going on and we can still predict with a degree of certainty what any animal may do next from an understanding of its current behaviour, because we appreciate that simple hard-wired responses may be modified by learning or internal motivational state and, crucially, we understand what those modifications are likely to be.

The next few pages

Of course all this oversimplifies; it makes it all sound very simple and mechanistic – and there are, in fact, all sorts of complications and refinements along the way, which we will explore in the next few chapters. But the basic truth is there. If you understand properly these various, essentially simple, processes and their influence in determining the behaviour of an animal, you should be able to predict exactly how it will behave in a given set of circumstances.

Such understanding, and the ability to predict, is of course only 'perfect' for very simple animals whose behaviour is mostly of simple reflex type, or, if it is modified by learning or motivation, is modified in exactly the same way in all individuals. As soon as one starts to get organisms whose individuals differ markedly in individual genetic make-up (and thus may have encoded different fixed action patterns) or differ markedly in experience and environment of upbringing (which means they will have had different learning opportunities) then such prediction becomes less certain. And that's the joy of trying to understand the behaviour of 'higher' animals where there is a far greater element of individuality. But even in such case, the understanding you will develop of the processes involved in building up observed behaviour patterns, while it may no longer enable you accurately to predict future behaviour at that level, still does allow you to understand and explain past behaviour once observed. And in effect, that's what this book is all about.

It's important to make one final point. While simple animals (with limited complexity of sense organs and nervous system) are only capable of simple behaviour, you shouldn't assume the converse: that as animals get more sophisticated so *all* their behaviour is of more complex form. Rather it is a cumulative thing. Simple animals (simple in neural complexity) can only cope with simple responses, but in more complex animals some behaviours remain of simple reflex type and only some become modified by influence of motivation, learning (or individual variation). Thus complex animals show simple as well as modified behaviours; (as noted earlier: most survival behaviours, such as escape responses, remain as simple unmodified reflexes even in animals capable of much more complex responses - and, interestingly, are controlled by the older parts of the brain). So while simple animals can *only* manage simple reflexes, the behaviour of more complicated animals may include simple reflexes as well as behaviour modified by motivation or learning.

As I emphasised at the beginning of this chapter, this simple overview is indeed simple – too simple. While true in essence and sufficient to present the general framework we have constructed, the "devil is in the detail". And over the next few chapters we will review each step of the way from the simplest reflexes, through a discussion of what aspects of their surroundings animals actually are aware of and thus what actually constitutes a stimulus, through analysis of learning and motivation, to refine and redevelop this model of complex behaviour.

2

Behavioural reflexes

Reflex mechanisms

We started last chapter's audacious tour of 'the whole of animal behaviour in 2½ pages' by noting that the most basic behaviour patterns could be considered as simple, almost physiological, reflexes between stimulus and response: that if exposed to a given stimulus for which it has a pre-programmed response, an animal will perform that response, whenever and wherever and as often as it suffers exposure to that stimulus. Such reflexes are common and, as we also remarked, may be observed in the behaviour of almost all animals, whatever their capacity for more complex response. Thus – even in complex organisms quite capable of more sophisticated behaviour – these simple behavioural reflexes are ideally suited to rapid survival responses such as escape from predator attack: responses which are straightforward, uncomplicated, have no need for flexibility and need to provide rapid reaction. For such responses, these simple reflexes are retained in organisms of the highest complexity. For organisms which cannot 'aspire' to learnt or motivationally-mediated behaviour, however – those which, physiologically, simply do not have the necessary circuitry – *all* behaviour must of course be controlled by such reflexes.

The essential thing about these reflexes is that they are absolutely fixed – and must be pre-programmed in the animal from birth. This itself may be achieved at two levels. First, the reflex may be no more than a purely physiological reflex arc, hard-wired into the animal's neural circuitry during development, so that if a particular sensory nerve ending is activated it is fixedly linked, through a series of fixed nerve connections, to a particular motor pathway and thus pre-programmed simply in terms of the embryological development of the organisation of the nervous system. Such a purely physiological reflex is involved, for example, in the removal of a finger when it accidentally touches the hotplate, let us imagine, of a cooker. Stimulation of heat-sensitive nerve endings in the finger tip triggers nerve impulses racing through a succession of sensory nerves which end, in mammals such as ourselves, in the spine. Within the spine, these sensory nerves are fixedly linked to motor nerves passing back out of the spine to activate muscles in the hand or forearm, withdrawing the finger from the heat. So rapid is this response and so fixed the circuitry that in fact the finger will be withdrawn from the source of heat before other nerve fibres passing up the spine to the brain 'inform' the brain of the situation. The finger is removed before you actually feel any pain and before, in effect, you even know about the problem in any conscious, cognitive sense.

Alternatively, the reflex may be a reflex more at the behavioural level. If, in sorting through incoming information from its sensory receptors, the organism happens to register a particular stimulus, then it sets in train a particular specific response, rather as an old-fashioned switchboard operator plugs in a particular extension when the number is asked for. In this case, the connection between stimulus and response, or sensory nerve and motor nerve, need not be permanently set up, but the link is made automatically once the stimulus is registered. Such behavioural reflexes are, of course, routed through some central nervous system, acting as central processor, and must be pre-programmed by some genetic behavioural blueprint of action and linked reaction. It's a bit like a computer programme which can be told:

If A , then do B,

But If C, then do D

In fact, such an analogy is not a bad one, for animals whose behaviour is completely ordered by such reflex action are, in effect, rigidly pre-programmed in this way and the 'brain' is little more than a well-programmed Central Processing Unit just as the one inside your computer.

Classic illustration of this second sort of reflex can be made by citing examples where it produces 'correct' but inappropriate responses. Large White (Cabbage White) butterflies[5] feed on the nectar from flowering cabbages and other crucifers. They are programmed to identify food plants simply by the colour blue. The behaviourist Robert Hinde noted that foraging butterflies of this species occasionally interrupted their feeding to flutter round the cover of his blue notebook, landing on the card and hopefully extending their probosces.

5 Scientific names for individual species mentioned in the text can be found on pages 177–178

In the breeding season, male three-spined sticklebacks (small freshwater fish) assume a bright breeding colouration of an electric blue back and bright red belly. They also become extremely territorial and aggressively attack and expel any intruding males, which they identify by the red of the belly. In experimental tanks on his laboratory windowsill, the Nobel prize-winner Niko Tinbergen noticed that all the male fish rushed to the window whenever a (red) post office van passed in the road outside. Inappropriate behaviour, perhaps – but red post office vans are less than common on the gravel bed of the small streams in which sticklebacks usually live! In such a context the colour red is almost always associated with an intruding male stickleback, so that the reflex, "attack red", is for the most part appropriate and effective. But such errors as this and of the butterflies attempting to feed on the recorder's notebook do serve to emphasise the essential inflexibility of the response.

The performance of such behavioural reflexes is constrained only by the limitations of the animal's sensory capacity in registering stimuli (discussed in Chapter 4) and the physiological capability of the nervous system in responding. Nerve-muscle systems are not instantaneous, though, and do have properties which affect their ability to respond and these limitations do impose some constraints on reflex behaviours of this kind. Indeed these same limitations impose constraints on all levels of behaviour including much more complex responses, because all depend on the same neuromuscular systems to perceive stimuli and organise a response. These limits, however, are more clearly apparent in the simplest reflexes, which are uncluttered by other sources of complexity. In more complex behaviours there are so many other additional sources of variation or limitation that those imposed by the simple operation of the nervous system are relatively small by comparison.

What, then, are these limitations of the nervous system? Reflexes, and more complex behaviours both show a **latency** in response: a small but perceptible time delay between the application of a stimulus and the appearance of some response. The latency, for example, between placing one's finger on the hot plate of a cooker and withdrawing it from the heat is somewhere between 60 and 200 milliseconds. This latency is known to be due simply to the time it takes for impulses to travel along nerve fibres (by electrical discharge), and across the junctions between one nerve fibre and the next (by diffusion of chemical transmitters). In reflexes, as above, it remains relatively short, but it is not surprising to learn that such delays may be greater in more complex behaviours where, in the chain between sensory receptors and the muscle or other motor system finally stimulated, there must often be dozens of nerve endings involved and many complex interconnections.

It has also been noted (and first reported by the physiologist Sir Charles Scott Sherrington) that some reflexes do not first appear at full strength. Yet, with no apparent change to the stimulus, the intensity of responses increases over a period of some seconds. This warm-up period seems to be due to the fact that, over time, a strong stimulus comes to elicit a response from more and more individual motor nerve fibres, producing a more pronounced response. This phenomenon is often referred to as **motor recruitment**, an entirely descriptive name since it explains that the behavioural observation of warm-up is due to a physiological cause: the fact that the stimulus at first elicits a response only in a small number of motor nerve fibres, but over time 'recruits' more motor fibres to carry the message, increasing the strength of signal to the target muscle involved in the response.

Finally, it is observed that after a period of stimulation, a response may wane in intensity, or switch off altogether. That this is not due simply to failure of the muscle activated, is clear from the fact that if the same muscle is stimulated artificially it will continue to respond for many hours. The tail-off, or cessation of response, is referred to as **fatigue** and appears to be due to an increase in the time taken for nerve cells to recover after repeated use. Transmission of nerve impulses through a nerve cell is due to a reversal of electrical charge across the cell membrane. After each impulse, intra-cellular 'pumps' transfer charged ions across the membrane to restore its original electrical differential and make it capable of carrying another impulse. Transmission of impulses from one nerve cell to another, as already noted, is dependent on the release of chemical transmitters, which must each time be 'cleared away' from the gap between adjacent nerves, ready for the next release of new transmitter substance.

Both these recovery processes take time, however infinitesimally small the period, and this time lapse is called the **refractory period**. It is suggested that the refractory period increases with repeated use of the same nerve fibres over a long time (it takes longer each time to set the nerve cells back up again for the next impulse) and that this may result in the observed effect of fatigue, at least where the phenomenon is apparent in reflex behaviour. In more complex behaviours, waning or complete disappearance of response may result from other factors such as habituation (page 31, 32) where in effect, an animal 'learns' to stop responding to an ever-present stimulus which is not actually affecting it.

Reflex behaviours

Having considered some of the properties of these simple reflexes in the abstract, let us now look at some examples of how they may be used in practice in an animal's behavioural repertoire – other than in the basic escape responses already mentioned. One of the commonest areas in which behaviour is controlled by reflexes alone is in orientation of the body and basic navigation in simple invertebrates. Changes in orientation may be achieved most simply by general, non-specific changes in an animal's speed of movement, or rate of turning (**kineses**) or in a slightly more complicated way through the animal taking up a particular orientation or direction in relation to the actual source of the stimulus (**taxes**).

Thus it is a fairly common observation that woodlice of various species tend to cluster in moist areas and rest in large numbers under damp logs or under bark.

Such behaviour is indeed often the subject of student practical classes in school, where woodlice are placed in a simple choice chamber with one end drier than the other; the woodlice all accumulate in the damper end. But the response has little to do with choice. In fact, it occurs purely as the result of a simple reflex response to humidity. Because the woodlice slow their speed of movement in moister conditions, they will, by definition, tend to accumulate in damper areas.

This response is simply achieved by a change in the rate of movement, but kineses can also operate to orient animals through an effect on the rate of turning. Thus the human

body louse (I like offbeat examples) maintains a straight course when external conditions are constant or becoming more favourable, but increases its rate of turning when conditions become less favourable. In the figure below, the bull's eye of the target represents the source of an aromatic food item; the contour rings around it form the boundaries of zones of decreasing intensity of chemical stimulation from that source, constant within each ring, dropping between that ring and the next.

It may be seen that by the simple expedient of continuing in a straight line as stimulation increases, or at least does not decrease, but turning as soon as the intensity of stimulation declines, the louse makes fairly quick progress to the source of the stimulation. And while this example moves the louse towards its food source, a similar response (where rate of turning increases as stimulation intensity increases) is what enables planarian worms to move away from bright light, where they are more at risk from predators.

These kinetic reflexes involve general changes to the animal's speed or direction of movement. In

taxes, the body is moved to take up a particular orientation with respect to the source of the stimulus. To continue our less than salubrious style of example: fly maggots, like planarian worms, move away from direct light. Maggots have a single light-sensitive organ at the front of the head; by swinging the head from side to side, they can make comparisons of the intensity of light on either side, and the animal may move accordingly towards the side of dimmer light. If an experimenter switches on an overhead light only when the head is swung in one direction (let us say to the right) then the intensity of stimulation received by the maggot when the head is swinging in this direction is always greater then when the head swings to the left. The animal can never balance the intensity of stimulation received on the two sides, so circles to the left. **Klinotaxis**, as this is called, is in fact relatively rarely used as a response to light, but exactly the same type of response is used by our old friend the planarian flatworm in response to the chemicals diffusing through water from a potential food source, enabling it to locate and move towards the food.

A different form of taxis involves the simultaneous stimulation of a pair of sense organs positioned on either side of the body and a response to unbalanced stimulation of the two, rather than a sequential comparison of stimuli received by a single sense organ swung in different directions. By having a sense organ on either side of the head like this, comparisons can be made at once, without the necessity of moving the head from side to side, but if an animal relying on such mechanisms has the receptor on one side destroyed it will circle continuously, since the intensity of stimulation of the paired organs can never be balanced.

With taxes like these, we are still in the realms of reflex behaviour: straight stimulus-response, but as we continue our explorations, you will see that we are moving more and more into the realms of more complicated mechanisms, involving 'analysis' of stimuli and acting on the conclusions of that analysis, rather than acting directly in response to the stimulus itself. And, in one final form of taxis, **mnemotaxis** – the orientation of animals in response to memorised landmarks – we find a reflex system controlling really quite advanced behaviour. Although in itself still a very simple type of orientation response, mnemotaxis is commonly used throughout the animal kingdom, even among more advanced organisms such as mammals and birds.

Perhaps the most elegant demonstration of the essence of this behaviour (and its reflex nature, as revealed by an inability to correct for 'error') comes from the very simple observations of Niko Tinbergen on the homing behaviour of the European beewolf - a delightful example because of its elegant simplicity. The beewolf is a solitary wasp whose reproductive strategy, like that of many other such wasps, is to dig a short nest burrow in sandy soil, provision it with living prey to sustain its grub, lay an egg in or on the paralysed prey and then seal the nest before flying off to repeat the exercise with the next egg. After digging a nest burrow it must fly off and capture some hapless animal prey; it paralyses it, but then must return with it to the nest burrow to bury it and lay its egg. And it finds its way back to the nest burrow using landmarks. These are memorised during a series of scouting flights in ever-increasing circles around the nest after it has finished digging the

nest burrow and before it flies off in search of prey. Tinbergen's elegant demonstration involved placing obvious, but artificial landmarks around the nest burrow before the wasp emerged, in the form of a simple ring of fir cones. After the wasp had emerged and flown off to hunt for prey, Tinbergen moved the ring of cones and re-assembled it a few metres to the side. On its return, the wasp flew directly to the centre of the (now displaced) ring of cones, not to the true nest entrance – and even though the nest was only a matter of a few feet away, it was unable to locate it.

Thus (unless some interfering experimenter intervened) the wasp was apparently able to fix the immediate location of its nest quite accurately with respect to immediate local landmarks – and able to repeat the same process for bigger and more distant landmarks as its orientation flight took it further and further from the nest, so that the whole local area became, so to speak, mapped out in its mind. As we shall see, such abilities are repeated in a whole variety of other insects, such as bees, and orientation in relation to memorised landmarks in this way is also fundamental to the local movements of most mammals and birds around a familiar range.

Many birds are able to navigate over huge distances during the course of long-distance, seasonal migrations (Arctic terns, for example, may migrate literally from pole to pole), often over unfamiliar territory, and in such case have been shown to navigate by celestial cues such as the position of the sun or stars, or using subtle changes in the earth's geomagnetic field (Chapter 15). Over shorter distances in the immediate local area of a familiar range, however, they too resort to orientation by known landmarks. Homing pigeons use a sun-compass or geomagnetic cues to travel from the unfamiliar point of their release to the general local area of the home loft, but switch to the use of memorised landmarks within the final ten-mile radius of home. And the popular press often carries stories of the remarkable abilities of pet dogs, whether lost or moved to a new home, to return over huge distances and many days of travel to their old home. Once again, the wider part of this journey over unfamiliar territory appears to be achieved by a form of true navigation using the position of the sun, or the night sky, but for the last few miles through more familiar territory, the dog again orients itself with respect to familiar landmarks.

Even here, then, while these more sophisticated animals are capable of using very complex systems of true navigation to guide them over unfamiliar ground, a pure, unrefined reflex behaviour, of orientation in relation to memorised landmarks, is fundamental to more immediate local movements.

Reflexes suddenly seem far from being merely 'simple' behaviours.

Linking responses and the co-ordination of more complex behaviours

The simple behavioural reflexes we have been discussing so far are, at their most basic, not only simple in circuitry, but also simple in the sorts of behaviours they may control. Essentially, one given stimulus releases an appropriate motor response. More complex behaviour patterns can be assembled, however, if a series of separate reflexes are joined together in sequence. Such sequences may be built up as a set of sequential responses to different features of the environment, but surprisingly commonly are internally-generated, such that the change in the animal's status brought about by its own last action, or even the actual performance of the action in itself, acts as the stimulus for the next action.

Thus some initial stimulus produces an initial response, which brings about some change in the animal's relationship with, or perception of, its surroundings, which in turn acts as the trigger for the next action, and so on. We have already seen an indication of this in the head-swinging maggot of an earlier example, which responds to a difference in light intensity on either side of the body by a small shift in body position. In that new body position, the next swing of the head brings an update on the change in light balance on either side and influences the next movement.

Or, in a slightly more complex sequence, the placing of new material into the nest cup during the construction of a nest by finches and other small birds involves a number of discrete motor actions. For each new stem placed into the nest, the bird performs four main actions:

- pulling and weaving a stem into position
- turning round in the nest for inspection
- sitting down in the unfinished nest and pressing the stem into position with the body
- wiping the new stem with the beak

Each of these actions acts as the initiator for the next (and the sequence is fixed and inflexible in form), so that the action of pulling and weaving the stem prompts the bird to turn around. Turning itself prompts the action of pressing into the nest cup with the body, while the sight of the new stem now properly positioned initiates bill-wiping. Although this behaviour is composed of a series of separate actions (actually somewhat simplified in this account), the links between these separate motor actions are so tight that the whole pattern is always performed as a single entity.

When such sequences of behaviour are patched together into a discrete behavioural unit, as in this 'placing' behaviour of a nesting chaffinch or, for example, a domestic canary, we are in fact observing the integration of a series of separate small actions into a single complex behavioural event – quite distinct from the one-stimulus:one response basis of simple reflexes. Such behavioural 'units' – sets of linked actions organised sequentially into a recognisable and fixed behavioural unit – are referred to as **Fixed Action Patterns** (page 3). These fixed action patterns differ from simple reflexes only in their complexity and the integration of a number of component individual actions. They are still inflexible in style and composition, so that once the sequence is started, the whole response is performed in exactly the same way, without variation, to its completion. And they are still genetically pre-programmed.

Other such fixed action patterns include the sequence of actions performed by a ground-nesting bird, such as a gull or a goose, in retrieving and rolling back into the nest an egg that has fallen outside the nest cup, or in the regurgitation of food and the feeding of chicks by parents in response to the begging of hungry youngsters.

This same idea – that a change in the animal's relationship with its environment, or at least in its perception of that environment, a change brought about by its immediate past actions, may act as the stimulus for the next action in some sequential pattern – is not merely limited to organisation of small individual actions into larger behavioural entities such as fixed action patterns. Exactly the same process may be involved on a larger scale in linking the individual fixed action patterns themselves into much more complex responses.

Thus, we talked of the placing of nest material into the nest cup by breeding canaries and how the separate actions of 'placing' were linked into a single action pattern. But 'placing' itself is only one of a series of actions in the overall behaviour of building a nest and the separate fixed action patterns of gathering nest material, bringing it back to the nest site, placing it in the nest, egg-laying and even incubation are linked together in much the same method of self-generating sequences of stimulus and response.

An egg outside of her nest stimulates a female greylag goose to stretch her neck out and hook the egg under her 'chin' before moving her head and neck to flick it back towards the nest. Each time she sees the egg outside the nest, the behaviour is repeated until, by trial and error, the egg is returned to the nest cup and so she can no longer see it as outside the nest

Nesting in fact consists of a number of different responses, each in itself a fixed action pattern. From studies in canaries by Robert Hinde and colleagues, we may define the sequence as

- Inspection and selection of a nest-site
- Gathering of nest material
- Carrying material to the nest-site
- Placing and weaving the nest material into the nest (we have already met this element)
- Egg-laying
- Incubation

Each pattern is linked to the others, and the switch between them is mediated by just the same sort of mechanism we have been discussing in explanation of the form of fixed action patterns themselves. Thus:

- The sensation of fullness of the beak from gathering, initiates the flight back to the nest

- The sight of the unfinished nest initiates placing, pulling and weaving
- As the diameter of the nest cup gets smaller and smaller, this in itself acts as a stimulus to switch from gathering grass to collecting feathers
- A finished nest cup is the stimulus to initiate egg-laying
- The presence of eggs in the nest cup acts as the stimulus to start incubation

Although there may be some flexibility in the system – and 'return loops', so that the sight of a smaller nest cup re-stimulates gathering, but of feathers rather than grass – the links between the separate patterns are quite specific and still essentially pure stimulus-response reflexes. A hen canary will switch off nest-building behaviour and commence incubation even in an unfinished nest if an experimenter sneaks a clutch of eggs into an uncompleted nest-cup. But such sequences of actions: where an animal's actual response to some original stimulus (or the altered situation created by such a response) acts as the stimulus for some other behavioural action, can lead to the development of really quite complicated behavioural responses, as this nesting sequence of breeding canaries amply demonstrates. Yet these are still, in their essence, sequences composed of simple, unmodified reflex actions. Indeed, they are characteristically rather stereotyped and rigid in form and performance.

Clearly, though, not all behavioural responses are so inflexible and we can identify many, more complex, behavioural sequences whose form is very variable. This is usually because, instead of only having strong, internal linkage between the separate actions, their performance is influenced by continued additional input from other external stimuli.

We may most easily visualise this if we may return to our computer analogy of the previous chapter. Instead of the simple

If A, then do B

of the pure reflex, or reflex series we often observe behaviour patterns more of the form:

If A and if also X, then do B or
If Either A, Or X, then do B
with additional "If, Then" choices at every stage.

Both devices introduce immediately a degree of flexibility of response - and these more flexible behaviour patterns, continuously dependent on reaction to additional stimuli during their execution, are of course much more common in complex behaviour than the all or nothing fixed responses we have so far discussed.

In the breeding behaviour of the three-spined stickleback (which we met on page 8), after a female fish has laid her eggs in a nest purpose-built by the male, he takes responsibility for the eggs, and performs a distinctive behaviour in the entrance to the nest tunnel. He swims forwards with his tail but back-pedalling, so to speak, with his pectoral fins, to push a gentle current of fresh water over the developing eggs, making sure that the water around them remains fully oxygenated to prevent them from being 'suffocated' by an accumulation of respiratory carbon dioxide in stale water around them. This fanning behaviour is itself a fixed response, but the intensity is controlled by continuous monitoring by the parent fish of carbon dioxide levels in the water. Thus, the fanning intensity increases as the eggs develop and if a batch of older and therefore more developed eggs (whose oxygen demand is higher) is substituted in the nest in place of the male's original clutch, fanning intensity rises at once in response to the new situation. [Interestingly enough in this example, the action of fanning is pre-programmed, although its intensity is not, and the male fish fans for the fixed length of time of the typical incubation period of eight days; in our experiment, therefore, where eggs of 6 days old are substituted for fresh eggs at the second day of incubation, while fanning intensity responds to heightened levels of carbon dioxide by increasing, fanning continues for a full eight days as it should have for the original batch of eggs, even though the substitute eggs have already hatched and the fry dispersed.]

So far we have talked of how simple reflexes may be built up into fixed actions patterns and how fixed action patterns or other reflexes may be co-ordinated and built up into quite complex and lengthy responses by sequential interaction of stimuli. Sometimes, of course,

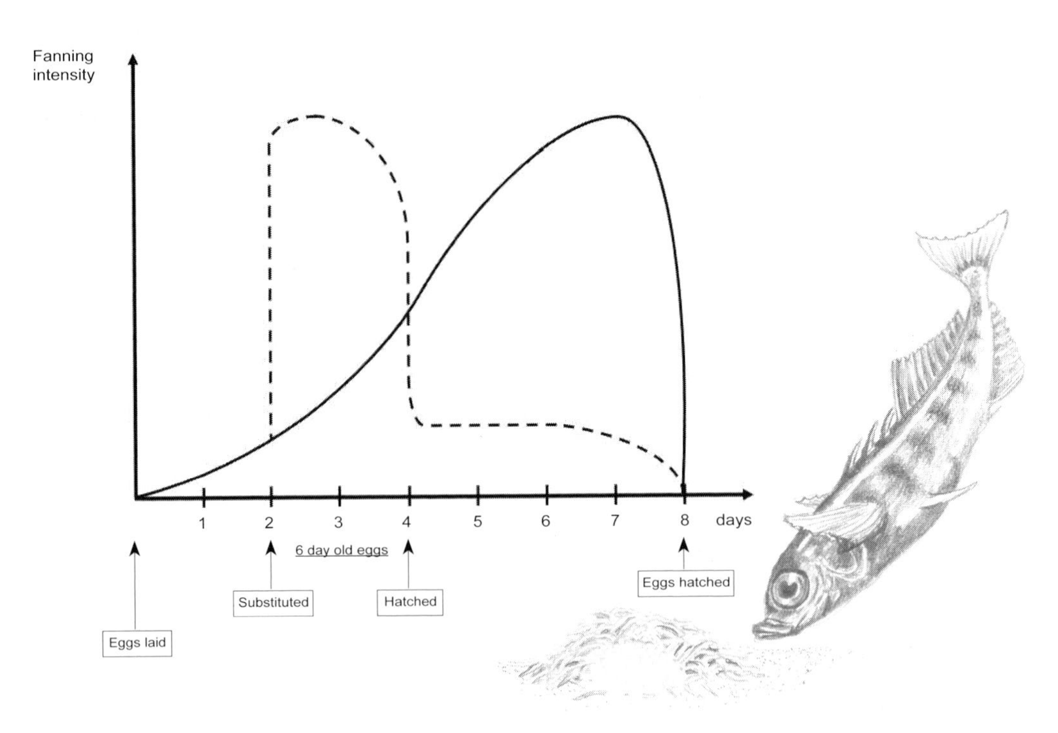

A male stickleback fans his pectoral fins to flush oxygenated water over the eggs in his nest. He continues this fanning behaviour for eight days, with the intensity of fanning responding to the levels of carbon dioxide he detects in the water. If a batch of six-day-old eggs are substituted for his original clutch on Day 2, fanning intensity immediately rises to respond to the oxygen demand of the older, more fully-developed eggs. Fanning intensity declines when these new eggs hatch two days later, but the male fish continues to fan at reduced intensity for the full eight days, even though no eggs remain within the nest those stimuli may not derive from an individual's own actions or from a passive external environment, but from the actions of other animals around it. While these, too, may not be entirely unpredictable, since they may in turn be a response to the first individual's own actions, it adds a new dimension to the development of behavioural patterns.

Many interactive social behaviours are in fact organised in this way and this, as so often, has practical functional value. If an animal, for example, were to start courting another individual of the wrong species, or one of the right species but in an unreceptive state, she will not respond to his initial advances. If his next actions are in turn dependent on the correct response to his opening gambit, the sequence will stop there, dead in its tracks, saving a great deal of wasted time and effort (and also minimising risk to the displayer since, while intent on courtship, his senses may not be so acute to detect risk, from perhaps an approaching predator). For illustration here, let us rewind the tape on the busily-fanning stickleback of our last example.

As we have already noted, at the onset of the breeding season, male sticklebacks assume a bright breeding colouration of electric blue back and red belly and establish small territories above the gravel beds of the small freshwater streams in which they live. Here the male scrapes out a shallow depression in the sand or gravel of the stream floor and builds above it a simple nest-tunnel of weeds. Rival males, recognised by their red bellies, are vigorously challenged and chased away. The male has a somewhat ambivalent attitude to visiting females, as although it's true that they do not have a red belly, they are nonetheless fish and intruders into his territory. And much of the courtship display has to do with reducing, or at least temporarily suppressing, his own aggressive impulses and inducing the female to lay eggs in his fine nest. In practice, gravid females, ready to lay eggs, have a swollen and distended belly which seems – as much as any lack of redness – to be the cue to their acceptance.

The male approaches the distinctively gravid female, assuring himself that she is indeed gravid and receptive. He then darts away again in the direction of the nest. If she does not follow him, he resumes his approach, indulging in a curious staccato zig-zag between the female and the nest which gives the courtship 'dance' its name. He repeats this to-and-fro first stage of the ritual until the female either leaves the territory or follows him to the nest. It is her behaviour, her action in following the male, which is the stimulus for him to switch from this zig-zag to taking up a curious, nose-down position at the entrance to the nest tunnel, 'pointing' to it with little forward and backward movements of the body.

Once again, the female may respond or she may not; if she shows an interest in the nest mouth, the little male swims through the tunnel as if urging her to follow. This behaviour, too, may be repeated over and over again until the female does indeed enter the nest. As soon as she does so, the male darts round to the other end of the nest and, placing his nose just above the female's tail, begins to nuzzle it in a series of rapid vibrations. This stimulates her to lay her eggs. She leaves the nest and the male immediately follows through to shed his sperm and fertilise the fresh eggs before emerging in his turn behind her. Now lacking a swollen belly, the female is recognised as an unwanted intruder and summarily chased away before the male returns to the nest to guard and aerate the eggs as we have already seen.

This whole courtship 'dance' is composed of a series of independent elements, each dependent on the previous behaviour of the partner. Thus the male continues to zig-zag until the female follows his 'lead' to the nest. He continues to 'point' until she investigates the nest mouth. That investigation prompts him to swim through the nest tunnel to 'encourage' her to follow, and it is indeed his action of swimming through the nest which does stimulate her to do the same. Her entry into the nest causes him to vibrate; his vibration causes her to shed eggs, and so it goes on in a tight, interwoven interaction of behaviour and response. At any stage in the whole ritual, the female may respond to the male's behaviours or she may not, and if she does not then the courtship ends, or reverts to an earlier stage in the 'dance', as the male goes back a step or two to produce the 'correct' response for the female's behaviour at that point. It is elegantly simple and extremely effective; as we have noted before, many survival behaviours are the better for their simplicity. In essence, still, the behaviour is a series of actions performed in response to a succession of stimuli from the animal's environment – stimuli emanating from its own internal state, from its physical surroundings and from the behaviour of other animals around it.

Receiving stimuli and analysing the incoming information

We have progressed some way, but we have been focusing on the response. We still have not really examined what we mean by these 'stimuli' from an animal's environment which control that response, have not examined the inputs to the system. In this chapter we will explore what different animals may be able to sense of the world around them, and how that perception actually controls the behavioural responses we may observe.

The stimulus configuration

The first thing we should note is that stimuli from an animal's environment may affect the performance of behaviour in a number of different ways: they can act to trigger the performance of a behaviour or they can alter that performance once the response has already been initiated, perhaps affecting its intensity or the 'target' to which it is addressed. Frequently, many of these different effects are under the control of one and the same stimulus – that is to say that the same stimulus that 'prompts' the performance of the behaviour in the first place may also control its intensity and direction. Equally frequently, however, different stimuli affect different aspects of the performance of a behaviour.

Indeed, Robert Hinde recognised two distinct and different categories of stimuli in the environment, noting that some cues were impermanent, present only occasionally or fleetingly in an animal's surroundings, while others were ever-present. He argued that the first type, impermanent cues, were much more likely to be important as triggers for initiating particular behaviours, while stimuli which are permanent features of the environment, simply because they are ever-present, are less likely to be used as triggers and their more usual role may be in channelling and directing a behaviour once initiated. As we will see, the fact that a cue may be ever-present in the environment does not necessarily mean that an animal is constantly aware of it, and its sensitivity even to ever-present stimuli may depend on its motivational state (Chapter 5), but Hinde's distinction is broadly quite helpful.

Characteristics of 'releasing' stimuli

Stimuli which are able to switch on the performance of a fixed action pattern, or more complex behaviour, generally consist of simple cues (odour or taste, colour, tone or pattern), whether singly or in some simple combination. These stimuli are often referred to in older texts as **releasers** or **releasing stimuli** since they act as the triggers to "release" the performance of a particular behaviour in appropriate circumstances. Some behaviours are released by observation of a single stimulus, while others require a combination of cues to be present, but the essence of such releaser situations is that they should be simple, something which defines a given situation precisely and incontrovertibly, yet with the greatest simplicity and clarity, so that it avoids confusion.

The aggressive response of the familiar European robin is stimulated by the red of a rival's breast (it literally does 'see red') and it is the colour alone which causes an aggressive reaction. Indeed, early experiments showed that a simple bunch of red feathers not even in the shape of a robin was sufficient to 'release' the response of attack. We have likewise seen that the colour red is sufficient to define the existence of a rival male in a breeding stickleback's home territory. The red need not be in the shape of a fish: the colour alone is sufficient to elicit an aggressive response, as we have seen in the misguided reaction of captive fish to a passing Post Office van; a mistake, surely, but in the fish's natural environment, 999 times out of 1,000, the colour red is sufficient on its own to define a rival. This is indeed a more general characteristic of releasers: that, in the main, most behaviours are triggered in their performance by a small part only of the entire stimulus configuration, rather than the whole – and are in effect triggered by relatively simple features of that stimulus 'whole'.

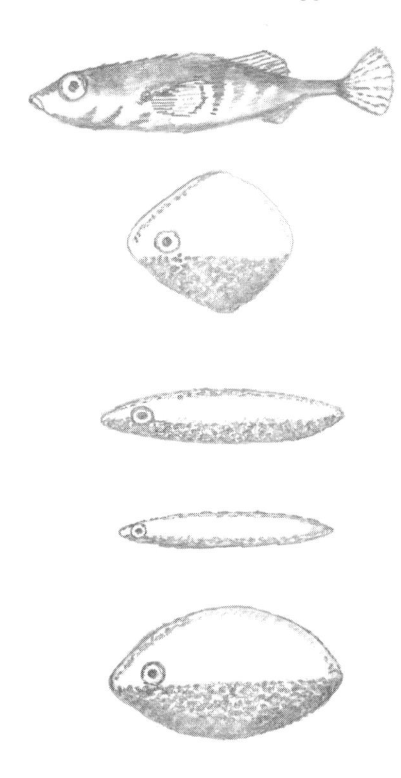

But, by the same token, since the animal often 'attends' to only one cue from the entire stimulus object, we can often elicit precisely which parts of the entire stimulus are important by the use of simple models. After all, our male robin reacted simply to a bunch of red feathers bearing no other resemblance to a bird. Careful presentation of appropriate models thus enables us to investigate the effects of varying one aspect of the stimulus object at a time. This kind of experiment is perhaps now rather old-fashioned, but the results have stood the test of time and the experiments themselves are elegant in their simplicity. We will content ourselves here with exploration of two classic examples.

Young gull chicks stimulate the regurgitation of food by parent gulls returning to the nest by pecking at the adults' beaks. This is in itself a classic releaser for a simple reflex behaviour of regurgitation (page 15). But what causes the young chick to peck at the parent's bill (and what directs the pecking?). What features of the parent's head and beak are 'necessary and sufficient' to define it absolutely and are thus appropriate to be used as simple releasers? Niko Tinbergen presented baby herring gulls with an array of simple models of gulls' heads, testing to see which elicited the greatest number of pecks. The shape of the head had virtually no effect, with heads of very irregular form eliciting nearly as many pecks as those more closely resembling an adult gull. What was important was the shape of the bill and the presence of a contrasting spot at the tip of the lower mandible; models which were long and thin, pointing downwards and with a red spot contrasting with the base colour near to the tip proved most effective at getting chicks to peck, clearly isolating these characteristics as the important features as far as the chicks were concerned.

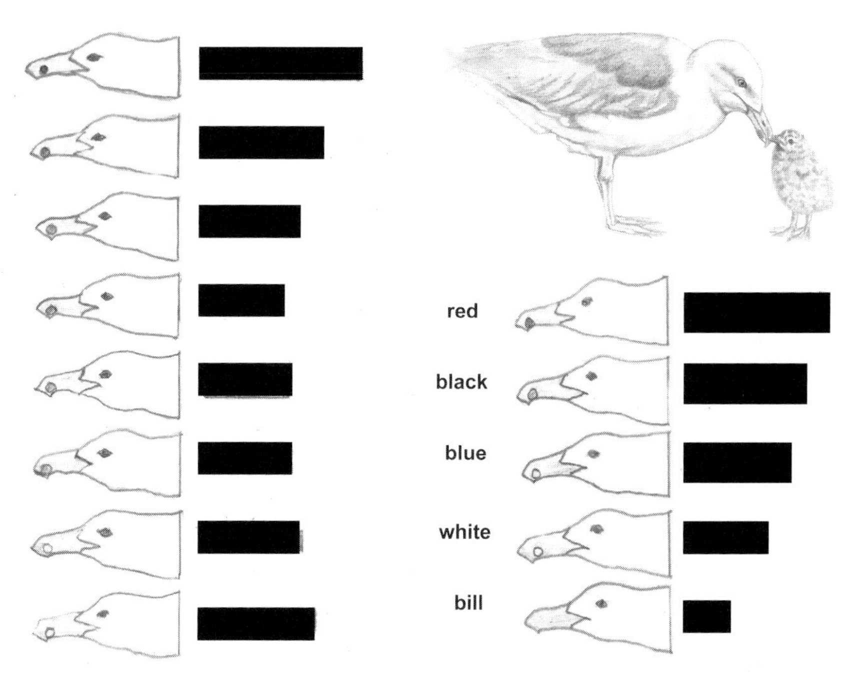

The shape of the gull's head has no effect on the willingness of a chick to peck at the bill of a dummy. What does, however, affect frequency of pecking is the presence of a spot at the bill tip and the degree of contrast between the intensity of that spot and the background colour of the rest of the bill. (Redrawn from experiments of Tinbergen and Perdeck, 1950).

A similar series of experiments sought to identify which features of a low-flying hawk overhead were significant in releasing the 'freezing' response in young chicks or goslings, whereby the young chicks crouch motionless until the predator has passed. The freezing response has tremendous survival value since the hawks themselves respond more strongly to any movement than to the shape of a stationary object; but chicks only freeze in this way when a predator is overhead and do not become motionless in the same way in response, for example, to a duck or a goose flying past. [It appears, in fact, that in the wild state this discrimination is not instinctive, but learnt: initially, chicks crouch at anything passing overhead but, as time goes on, they learn to pay no attention to common objects such as adult ducks and geese flying overhead or falling leaves, and continue to crouch only in response to unfamiliar objects which, because of their unfamiliarity, they have not learnt to ignore.]

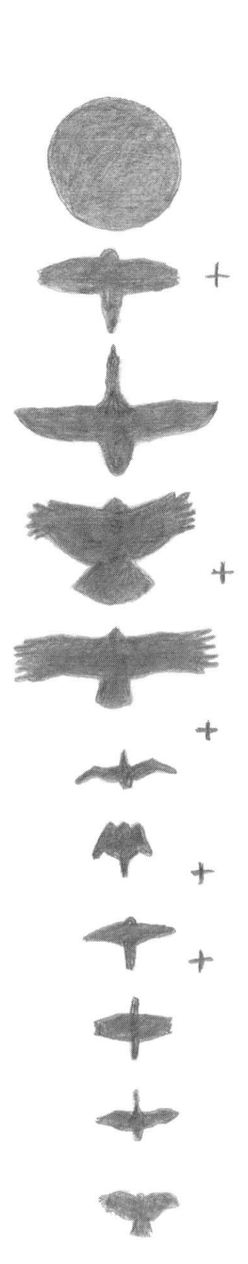

Tinbergen 'flew' a variety of different cut-out model shapes over pens of chicks and measured their responses both to extremely life-like models of various kinds and those becoming more and more stylised in form. He noted that models which elicited the freezing response all had broad 'wings' sticking out on either side and a 'neck' which was shorter than the 'tail'.

A single stylised model which could be considered to represent a hawk or a goose, depending on the direction in which it is moving (defining head and tail) only elicited the 'freezing' response when flown in the direction which gave it a short neck.

Stimulus summation

Even though there is a functional advantage in reducing the 'stimulus characteristics' of any given situation to a discrete few, readily recognisable features which are in themselves sufficient to define that situation absolutely, relatively few behaviour patterns are triggered by a single cue. In most cases, performance of the behaviour is in response to the simultaneous co-occurrence of more than one cue (although still a small number). In this case, we may observe that the 'goodness of fit' of one characteristic aspect of the stimulus set may compensate for some 'lack' in another aspect of the same object in a phenomenon often referred to as stimulus summation.

The retrieval of an egg which has rolled out of the nest depends, in the herring gull, on the size, base colour and degree of spotting of the egg. A larger egg is more effective in eliciting the response than a small one, a green egg (the normal background colour) is more effective than a brown one and a spotty egg is better at triggering the retrieval response than is one of uniform colour which is otherwise the same in all other respects. A model that combines all cues is most effective of all, but any missing cue can be 'compensated for' by enhancement of another: a small egg, for example, may be retrieved as readily as a normal egg if it happens to be more spotty.

Supernormal stimuli

This fact that the effects of different components of the stimulus situation or stimulus object may prove additive in their ability to provoke a response, and that 'exaggeration' of one property or another may compensate to some extent for deficiencies in some other property, leads to one additional oddity. Since it appears that the ability to trigger a response increases with increasing intensity of the appropriate key stimuli, there is a capacity for the development of so-called supernormal stimuli – "larger than life", which release an appropriate (or even sometimes an inappropriate) response with heightened intensity, simply because their stimulus effect is particularly intense.

Once again, this is most obviously apparent with models, where some relevant feature may be manipulated to exaggerate it to extremes, accompanied by a particularly intense performance of the linked behaviour. Thus if ground-nesting birds such as herring gulls, greylag geese or oystercatchers are presented with dummy eggs many times larger than their own, they will retrieve these into the nest (page 16) or incubate them in preference to their own.

This potential for superstimulation by supernormal stimuli is mentioned here in large part to give further clear illustration of the existence of the sort of stimulus summation we have been discussing, and to demonstrate particularly clearly that the increased intensity of even a fixed stimulus may increase the likelihood of a response. But in fact there are situations where these supernormal stimuli may affect behaviour in the real, natural world. Sometimes this behaviour, too, may be inappropriate. To the casual glance of the human eye, dunlin in their summer breeding plumage quite closely resemble smaller versions of golden plover; on the frozen tundras of the north, I have myself quite often observed clusters of ardent male dunlin, in the frenetic excitement of their courtship, in hot pursuit of a female golden plover.

An oystercatcher incubates a super-sized egg

Sometimes, too, the 'mistake' is deliberately contrived and exploited. The egg of the European cuckoo resembles closely the eggs of the small passerine birds in whose nests it is laid, and thus presents a supernormal stimulus to incubation – perhaps ensuring its acceptance; subsequently, the large yellow gape of the emerging nestling provides a supernormal stimulus for its foster parents in its demands to be fed, ensuring that it obtains the large bulk of food returned to the nest before it evicts its nest mates, and thereafter ensuring almost obsessive, addictive provisioning by its unfortunate foster parents.

Interactions between external and internal stimuli

In these last few paragraphs we have been discussing how various behaviour patterns may be triggered or released by single 'key' stimuli, or single aspects of some stimulus object, while other behaviours may require the presence of a number of stimuli acting together to trigger their performance. In such a context it is important to note that such interaction is not necessarily always between sets of external stimuli, but may be between stimuli from both the external and the internal environment. In the first few pages of the Introduction, I noted that simple reflexes might be modified by considerations of the animal's internal state (there is no point in continuing to drink if your stomach is already full of water) and it is appropriate to consider here, as a final point, that interactions between stimuli, and expression of behaviour in response to the summation of a number of independent cues, may in the same way involve some interaction between external cues and stimuli derived from the animal's own internal environment. Indeed, this interaction is essential.

At any one time an animal must in practice be receiving hundreds and hundreds of different stimuli from its environment. How does it select which to respond to? Further, amongst those hundreds of stimuli impinging upon it, it must often be exposed to the appropriate stimuli for the release of a number of different (separate) behaviour patterns, yet it only ever does one thing at a time. How does it select which behaviour to perform of the many possible alternatives for which appropriate stimuli are present?

The interaction between internal and external stimuli is crucial to the resolution of this second question: which of a variety of possible behaviours for which the correct external stimuli are present should an animal 'choose' to perform? In effect, we are not yet in the realms of choice, for an animal will only perform that behaviour for which the correct external and internal stimuli are present, and an animal which has already drunk its fill is receiving no internal nudges about dehydration, even though the correct external cues to prompt drinking may all be present and correct. Even the 'choice' between those behaviours for which both the correct external stimuli and appropriate internal stimuli are present is decided for us. In the phenomenon of stimulus summation, we have seen that the performance of any behaviour is affected by the summed intensity of stimulation. So it is that the behaviour for which the stimuli, both internal and external, are not only all present, but at their most intense, which will finally be performed.

And what of the earlier question? Given that an animal must at any one time be bombarded with many hundreds of different sensory inputs from its external (and internal) environment, how does it select out the relevant cues to which to respond? This is one huge advantage in having specific 'releasers' for different behaviours (page 22) and a releaser configuration slimmed down to the barest few cues. This characterisation of a few simple features, sufficient adequately to identify the situation or define a given object, is of tremendous benefit in enabling animals to filter out what are indeed relevant cues from what otherwise risks an information overload. But there is more to it than that, and animals are commonly able to filter incoming information as it arrives and select out what is relevant to the given situation; some of these filters are fixed, but others can be altered in changing circumstances or with changing need.

Filtering incoming information

Incoming stimuli from the environment may in practice be screened or filtered by two systems within the body: they may be filtered directly within the sense organs themselves, or within some central processor. Because of their relative positions in the animal's relationship to the environment, the effects of these two separate processes are commonly referred to as **peripheral filtering** and **central filtering**.

Some filtering of information within the sense organs occurs simply as a consequence of limitations of the sensory capacity or discrimination of those organs. If their sophistication or range of operation are restricted, then – very simply – a whole potential array of incoming stimuli are eliminated from consideration because the animal simply does not perceive them. But in addition, there would appear to be the potential for a more active filtering process in the sense organs (or sensory nerve pathways) such that, at times, stimuli from the internal or external environments are perceived and registered, yet the information is not passed on to the central nervous system.

It is perhaps more than obvious that the number of potential stimuli any animal may receive from its environment is ultimately restricted by the range of its sensory capacities – but it is an important point nonetheless. Colour vision, for example, is somewhat erratically distributed in the animal kingdom. Quite simply, an animal which cannot perceive colour cannot respond to colour cues and must react only, perhaps, to changes in intensity. Even those animals which can perceive colour may do so over different bands of the spectrum; thus while humans cannot directly perceive light in the ultraviolet scale, many insects have extremely good perception of light in this wavelength. Similarly, different animal species may have different ranges of sound frequency over which they are able to 'hear'. Visual or auditory information transmitted at wavelengths or frequencies outside the animal's sensory range will simply fail to be registered. And a very great deal of 'de-cluttering' of incoming information, reducing the input to a more manageable level, is the direct consequence of such sensory limitation.

Following on from the discovery that there were, within the optic nerve of a frog, bundles of nerve fibres reacting to light in three different ways (light being on, light being off and the change of state from on to off), Lettvin and co-workers discovered that the

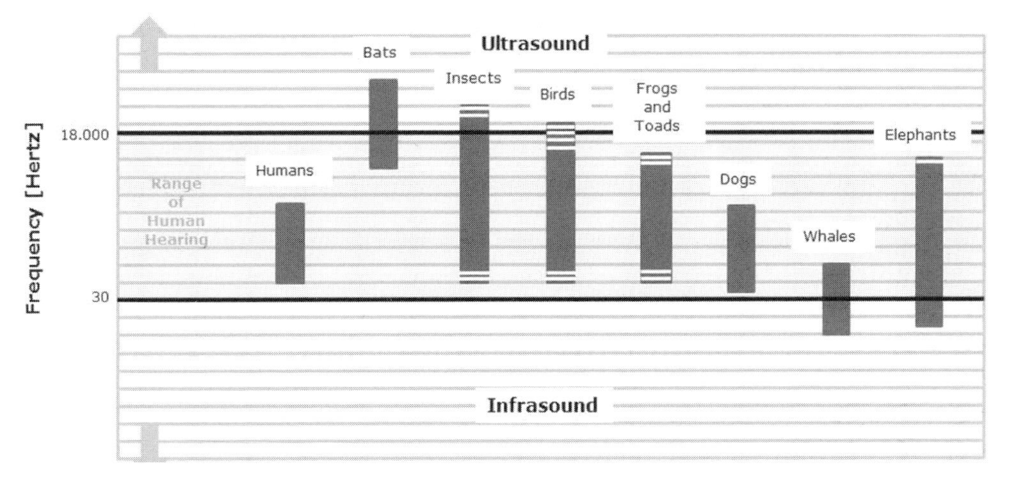

Frequency of sound transmission from some different animals, plotted against the range of human hearing. It is clear that our auditory limitations filter out from us a range of available information (Redrawn from information in Payne, K. (1989) Elephant Talk. *National Geographic* 176: 267)

whole visual 'machinery' of the frog's eye was restricted to perception of only four distinct 'events'. In an article called 'What the frog's eye tells the frog's brain', they reported that frogs could detect sustained contrast: there were sensory cells which began to transmit impulses in reaction to contrast within the visual field – sensory systems corresponding in effect to 'light on' fibres. Other receptors, generating 'light off' impulses, appeared able to detect only net dimming within the visual field. Frogs, it appeared, could also detect a moving edge, with curvature larger than the visual field, corresponding to a large object moving across the animal's field of vision and not wholly contained within it. Finally, Lettvin and his colleagues found what they called a 'bug detector': a set of sensory cells responding to a small object passed into the visual field and wholly contained within it. These do not respond for small objects already present in the field of view, but only those moving into it or moving across it.

If this is all that frogs can "see" it perhaps explains why they do not respond to stationary prey items, but attempt to capture prey only when it is moving: not because of some curious behavioural quirk, but simply because that is all that they are able to see. Such moving prey will be registered by the bug detector, while larger animals – potential predators of the frog itself – will be logged by the effect of net dimming of light and movement across the visual field of an object larger than the field of vision itself (moving edge detectors). Simple but functional, and apparently sufficient, since frogs continue as a successful group.

Not all such peripheral filtering, however, is due simply to limited capacity of the sense organs themselves. In some instances, the stimulus may be perceived by the sense organs, but the information received is not necessarily passed on. You will remember that when we discussed the characteristic of stimuli, we noted that we could define these in two main categories: those which were transient or ephemeral, and those which were ever-present. And we noted that transient cues were more likely to act as triggers for particular behavioural responses.

The reason for this is that ever-present stimuli, even if registered by the sense organs, are unlikely ever to reach the central nervous system or central processing unit. This may in part be a consequence of simple physiological processes such as fatigue, or the ever-increasing recovery time required to regenerate nerve cells for repeated transmission (page 9), but commonly it is the consequence of a simple (though unconscious) learning process called **habituation**. When any animal is repeatedly or continuously exposed to a particular stimulus from its environment, which turns out to be totally irrelevant, it 'learns' not to respond to it, and while this may be in larger part more a function of some central filter which may discard such irrelevant information, there is also some evidence for actual active suppression of sense organs or sensory nerve pathways in registering the object in the first place.

One experimental demonstration of this phenomenon which always delighted me involved measuring nerve impulses from the *nucleus cochlearis* of a cat enclosed in a dimly-lit room, empty apart from the cat itself and a ticking metronome. The *nucleus cochlearis* is a sensory ganglion in the auditory pathway from ear to brain. In such a set-up, regular pulses of activity in the *nucleus cochlearis* can be recorded, corresponding to the cat's perception of each click of the metronome. When a mouse is also released into the room, however, the electrodes no longer record any impulses from the auditory nerves in response to the clicking of the metronome.

By a shift in attention, the cat has actually 'switched off' its hearing of the metronome and not only does not attend to it, but actually does not hear it at all. This experimental

demonstration by Desmedt and Monaco of what is nicely referred to as 'afferent throttling'[6] is now supported by more recent physiological evidence which has established a known feedback mechanism from the central nervous system which is able to inhibit transmission of information from the *nucleus cochlearis*.

But despite the attentions of filtering processes in the sense organs and sensory nerve pathways, there are still a lot of stimuli which get through. Even if we know a great deal about the animal's sensory capacity, we still cannot be sure that all the stimuli which are registered and transmitted are necessarily behaviourally relevant. There are obviously other sorting mechanisms to select the immediately important cues. At the simplest level, the key releaser stimuli for simple behaviours may be pre-programmed genetically, so that the brain (or its equivalent) always recognises them at once: perhaps cross-checking incoming stimuli against some internal 'reference library' of important cues, and discarding those received stimuli which do not find a match. This 'library' may be composed initially of genetically programmed cues, but may subsequently be added to by learning, and the development of specific, learnt search images.

There may also be – as we have already mentioned – clear effects of selective attention: there may be heightened sensitivity to some cues rather than others, related to and perhaps induced by a particular 'internal state', such that the cues relevant to performance of behaviours currently of high priority become in some way more 'apparent'; or, in other cases, 'downgrading' the importance of cues temporarily or permanently of less significance (habituation).

The existence of some sort of filter of the 'releaser match' type was first postulated as a necessity from early observations of the sorts of simple reflex behaviour and fixed action patterns we have considered in earlier chapters. We have already noted that such actions are commonly released, or triggered, not by the natural stimulus object in its entirety, but by a few key features only (such that simple models, presenting only one or two features of the true stimulus object, may be sufficient to provoke a full and complete response).

6 a lovely term for the suppression of impulses in the afferent nervous system, which transfers information from the sense organs to the central nervous system

The obvious conclusion is that these crucial cues are those which are encoded in some pre-programmed library of such releasers. This makes intuitive sense, at least in relation to behaviours which are inborn, pre-programmed and essentially simple reflexes. For such behaviours it is appropriate to have some simple, unequivocal cue for their release; the definition of a few key features, necessary and sufficient to define the situation in which the response is appropriate or requisite, and then the encoding of those in some central processor to be matched (or preferentially selected) from a range of incoming stimuli, seems entirely appropriate – and there is indeed good evidence for such innate filtering mechanisms.

It is also makes intuitive sense that there may be ways in which changes in the animal's internal state, or changes in selective attention, might heighten or suppress sensitivity to certain of these (or other) cues, although the physiological mechanism is less apparent. Once again, we have good evidence that foraging animals which are seeking a certain type of prey improve their performance over time (even within any one feeding bout) and that having once captured one or two prey items and 'got their eye in', so to speak, they show improved efficiency in detecting later items of the same type. This development of a 'search image' is indeed part of common experience, even in humans. If you are searching for a certain type of object, particularly if it is against a complex background in which the object of search is well camouflaged, it is often extremely difficult to find the first. But once you have, so to speak, locked on to the pattern – the image of what that object looks like against that background – it proves much easier to locate the next object and the next.

We can equally readily demonstrate an increased insensitivity to irrelevant stimuli, which we have already described. Animals fail to register many ever-present stimuli which have no behavioural relevance. They may also become temporarily less sensitive to (or 'learn to ignore') stimuli which are irrelevant at the current moment, even though these same stimuli may be important at other times. Small fish such as our stickleback, or aquarium pets like guppies, offer a rapid escape reaction when presented with a shadow overhead (a potential predator). In response to the same 'net dimming' as Lettvin's frogs, they stop immediately what they are doing and dash for cover. If the shadow and the apparent danger passes, they come out from hiding and resume their previous activity. If a human experimenter passes a shadow over a fish tank time and time again, yet this is never followed up by any actual attack, reinforcing the 'threat', the fish begin to show a reduced responsiveness. To begin with, they do indeed dash away and take cover, but with repeated stimulation, the time interval spent before they return and resume their previous activity gets shorter and shorter on each occasion. Finally, they 'ignore' the shadow stimulus altogether and simply fail to respond at all.

This learning (that on this occasion the shadow has no relevance) is accompanied by a decreased responsiveness to the same stimulus – even to the extent that it may not be registered at all. In our example, such habituation is temporary only; if the fish are given a rest from repeated presentations of a shadow stimulus for a number of hours, and then exposed once more to an overhead shadow, the response will be found to be back in full intensity. After all, this time, it might be a real predator. Equally, even when fully habituated after repeated stimulation by shadow alone, having learnt that it was not accompanied by any attack, if the experimenter then passes a shadow over the tank and this time plunges a hand into the water at the same time, the fish will dart for cover once more – and will do so next time even in response to the shadow alone. The learnt 'observation' that passing shadows were irrelevant has now been superseded and reversed by observation that, yes, shadows are indeed still sometimes accompanied by attack so should not be 'ignored'; full sensitivity is restored. In other cases, when ever-present stimuli are never found to be relevant, habituation may itself be established longer-term, so that the decrease in sensitivity of the 'central filter' is fixed.

In these discussions of stimulus filtering, one final nicety remains. If many reflex behaviours may be released in response to only a few key features of a given object or stimulus situation, what determines the evolutionary 'choice' of releaser stimulus in the first place? Frequently we find, in a kind of self-fulfilling loop, that the key features selected are 'chosen' because they are ones to which the animal's sensory organs are particularly finely tuned in the first place.

Capranica studied the vocal response of bullfrogs to the mating call of another male. For the response, he discovered, the eliciting call must contain sufficient energy at around 200 cycles per second and also sufficient energy in the range between 1400 and 1500 cps. It must also have no components in the mid-frequency range (between 500 and 700 cps) with amplitude greater than that of the peak in the low frequency (200 cps) region. Anatomical and electrophysiological studies suggest that these frequency ranges are actually those best fitted to the auditory mechanism of the frogs themselves, which have been found to have two types of auditory sense cells: one sensitive to frequencies between 1000–2000 cps and the other sensitive to frequencies up to 700 cps. Interestingly, the calls of juvenile frogs have energies peaking in the range 700–1200 cps, a range where stimuli are either not detectable by adults, or actively inhibitory on the aggressive response: an elegant twist in the tail.

5

Motivation and its capacity to modify simple response patterns

At its simplest level, as we have suggested, any animal's behaviour may be treated as a simple reflex in much the same way as any physiological reflex: pure stimulus/response – whether this applies to individual simple actions or a linked series of actions within some stereotyped Fixed Action Pattern. More complex behaviours may depend on a whole series of stimuli interacting in various ways, and this interaction may be between a complex of purely external stimuli or, as we discussed in the last chapter, may involve some interplay between external stimuli and so-called internal stimuli: cues, if you like, from the animal's internal physiological and emotional state.

One result of this is that, as internal conditions vary, we may at different times observe different responses to exactly the same external stimulus situation. We have already suggested that this is one of the ways in which animals may 'choose' which behaviour to perform at any one time, when faced with the correct external stimuli for a number of different alternatives: that is, to perform the behaviour for which the appropriate internal stimuli are also present (and, from amongst those, the ones for which the internal stimuli are most 'intense'). We took the example of an animal presented with the correct external stimuli to promote drinking: perhaps, to a dog, presentation of a bowl of water (page 4).

Whether or not the dog will actually respond by drinking will depend on whether or not it is 'thirsty' or has recently drunk its fill, assessed in practice by internal sense organs in the carotid arteries monitoring the osmotic potential of the blood (and thus the level of water deficit within the body). If the dilution rate of the blood is adequate, the dog will not respond, even to the external stimuli presented by the water bowl, by drinking; in contrast, if the blood fluids are somewhat more concentrated than is ideal, the dog may indeed respond to the external presentation of the water bowl by beginning to drink, with the probability that it will actually drink increasing steadily the greater the level of dehydration detected in the internal body state.

The effects of internal stimuli may be direct in this way (simply as one component of the overall stimulus set for consideration) or may operate by changing in some way the animal's selective attention for different cues. When grayling butterflies are sexually activated and in search of mates, they pay no attention to colour cues. This is not due to a lack of ability to perceive colour, because when foraging normally for nectar, like other butterfly species, they have a tendency to respond most strongly to colours at the blue end of the spectrum. It is, rather, a change in responsiveness, akin to the stimulus filtering processes we were talking about in the last chapter, in this case brought about by simple changes in the internal level of sex hormones.

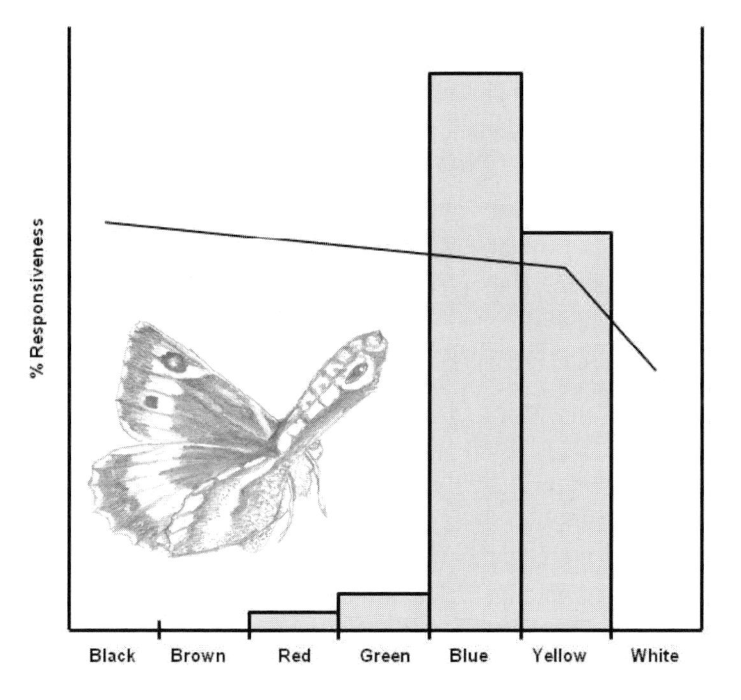

When foraging for nectar, grayling butterflies distinguish colour and respond most intensely to colours at the blue end of the spectrum (shaded bars). When searching for mates, males pay no attention to colour cues (solid line)

So, the change in responsiveness to a constant external stimulus situation may be mediated in a variety of different ways, but however it is actually expressed, we define the sum of the different internal stimuli which may affect an animal's behaviour or behavioural choice as defining its '**motivational state**'.

Various things may influence motivational state. The physiological fact of a water deficit affects the probability of drinking behaviour; and in a more general expression of this same observation, the status of a whole variety of different physiological conditions may influence the likelihood of performance of a particular behaviour: conditions which we may more readily refer to as hunger, thirst and so on. Hormones may have an effect: it is the change in the level of sex hormones in the body that effects the change in responsiveness to colour cues in our grayling butterfly and it is the release of floods of adrenalin in response to risk or potential threat which influences the performance of escape behaviours or aggression. All these in effect may be considered actual stimuli derived from the animal's internal state. But the first thing to emphasise is that these internal stimuli only contribute additional cues to the entire stimulus situation effecting behaviour. Internal stimuli cannot produce behaviour on their own in the absence of the appropriate external stimuli; rather they affect an animal's responsiveness to whatever external stimuli are present. In other words, those features of the animal's internal environment contributing to its motivational state can affect the likelihood of performance of any particular behaviour only in response to the correct external stimuli.

One of the most powerful contributors to the overall internal state is what used to be called 'drive'. This, if you like, is the internal pressure that increases the probability of performance of a particular behaviour pattern, the behavioural "oomph" that responds to a particular set of external stimuli if the motivational state is right. The word is now somehow considered politically incorrect, for it offers an oversimplification for a complicated suite of factors which may involve rather different physiological mechanisms in different specific instances – and because there is, in physiological terms, no direct counterpart for this nebulous 'behavioural energy'. In addition, the term 'drive' itself implies some internal urge towards performance of one behaviour or another, while in practice the expression of behaviour is a passive response to the co-occurrence of a set of appropriate internal and external stimuli. 'Drive' indeed merely represents some rather abstract reflection of the strength of the internal stimuli appropriate to the performance of each behaviour, an integrated summation of the combined intensity of those various specific internal cues.

But, with those reservations, the term may serve us as a useful device here for precisely that same reason: as a general cover-all term for that same disparate group of internal factors which may alter the probability of an animal performing any given behaviour.

It is intuitive common sense that while the probability of performance of some behaviours (such as escaping in the face of attack) is absolute and largely independent of an animal's internal state, the likelihood of performance of other behaviours may increase with the length of time since they were last performed (so drinking, once again, or feeding, will show increased 'importance' the longer an animal has been without food and water). It was this demonstrable build-up of probability, this increase in the 'motivational status' of particular behaviours with time since their last performance, that early behaviourists could see and could incorporate into their image of a behavioural 'drive': an increase in the sensitivity to appropriate cues and an increase in the likelihood of the performance of that behaviour again in the future, the longer the time elapsed since its last performance. These early ethologists assumed that 'drive', as they considered it, was motivation and that the two were merely reflections of one and the same thing. These days we accept that this behavioural 'drive' is only one element contributing to the entire motivational state as a whole and that even though the drive for some particular behaviour may be extremely high, yet other considerations (hormonal state, for example) may render the behaviour of comparatively low probability overall.

Clearly both this 'drive' and the build-up in probability of performance of a behaviour after a period of non-performance have a precise physiological basis, which we will explore in just a moment; but before we do so, it is helpful for the moment to consider it in these rather more general terms, to see by analogy how it might affect behaviour or an animal's response to external stimuli. Indeed one of the most easily visualised analogies was developed by Konrad Lorenz in terms of a simple hydraulic model, so familiar in domestic terms that it came to be dubbed the "Flushing Lavatory".

In Lorenz's imagery, the probability of performance of any behaviour, in relation to encounter of the correct external stimuli for performance, was influenced by the level or intensity of a series of facets of the animal's internal state. These he summarised by the concept of a simple tank of water (or 'drive') linked to each behaviour, whose 'head of pressure' related to the intensity of internal cues for its performance. For some behaviours, this likelihood of performance would be a constant. In other cases, it would change as time elapsed since the last performance of that same behaviour. Lorenz accommodated this in his analogy by having each tank fill up slowly – with the rate of filling determined by the behavioural importance of that particular behaviour, or the required frequency of repetition: a faster fill rate for behaviours needing frequent repetition, a slower rate for behaviours needing to be performed only infrequently.

In response to exposure to the correct external stimuli (envisaged as a lock and key valve) the cistern could be emptied. Once started, the behaviour would be fully expressed to exhaustion of the 'need' as an all or nothing behavioural response. The tank would be emptied in a single flush, and at once we see the approximation to the domestic lavatory. We have already noted however that, especially when the motivation for a particular behaviour is high, that behaviour may be performed even in response to rather incomplete external stimuli. Lorenz provided for this by allowing his lock-and-key valve to be spring-loaded: thus if the head of pressure in the cistern were high, the behaviour might be expressed in response to incomplete stimuli which would not normally be sufficient to trigger it.

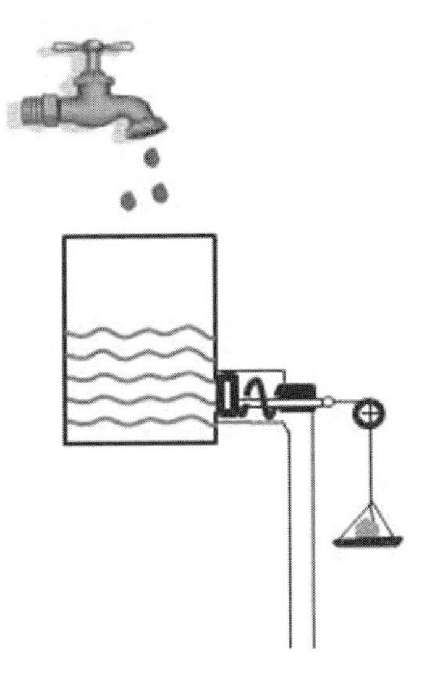

Lorenz's conceptual representation of the build-up and release of 'drive'. Energy builds up in the header tank at a rate depending on the importance of the given behaviour, with the level in the cistern reflecting time since the behaviour was last performed. The tank is flushed in a single flush in response to presentation of the correct stimulus. Where the motivation for a particular behaviour is high, the fact that the outlet valve is spring-loaded means that the behaviour may be performed even in response to rather incomplete external stimuli which would not normally be sufficient to trigger it

Let us now extend this simple imagery to consider a series of different behaviours, each with their own cistern linked to its specific releasing stimuli, each filling at different rates (dependent, as before, on the 'repetition frequency' required of that particular behaviour) – yet each inevitably needing to express its particular behaviour through the same nerve and muscle systems of a single animal body.

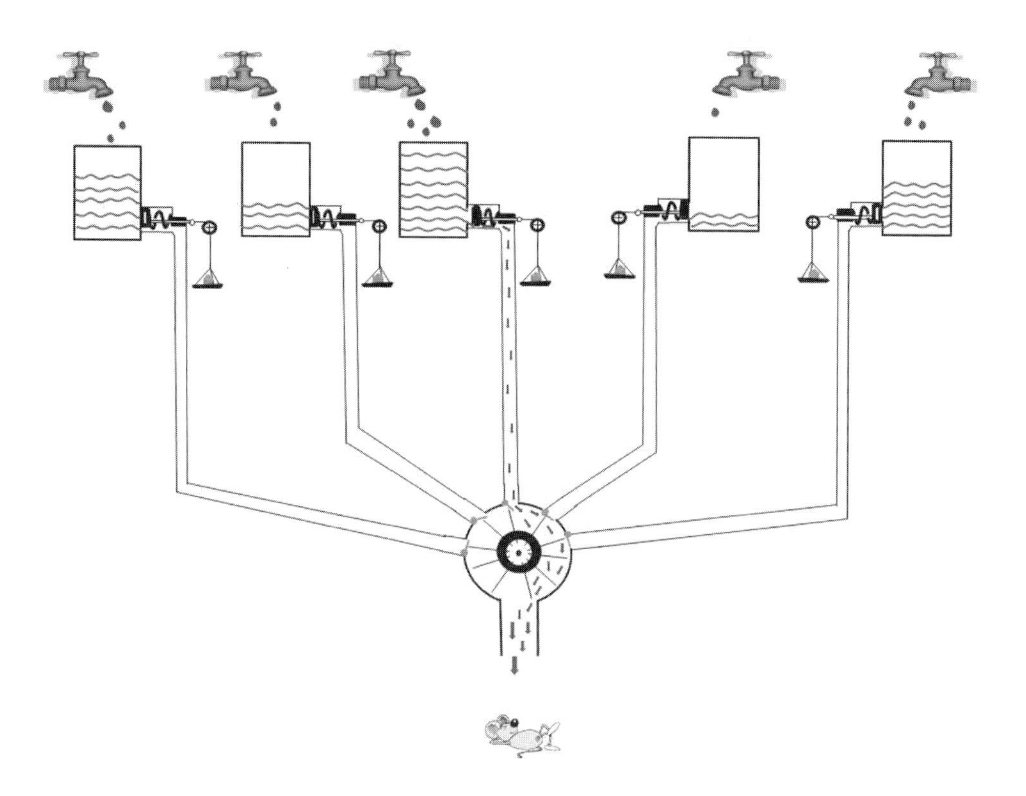

Since the animal can perform only one behaviour at any one time, there must be some mutual inhibition, in that the expression of one behaviour temporarily suppresses the potential for expression for all others. We may incorporate this in our hydraulic model with some common valve (X). If this valve, like the behaviour-specific valves, is also in some way spring-loaded, we can consider at least a hydraulic system where, of all behaviours for which the correct external stimuli are present from the environment, the one which will actually gain expression is the one with the highest water pressure in its tank (related both to absolute time since last performance, and a rate of fill related to the behavioural importance of that response).

Once again, we must reiterate that this is an image, a model only and we do not seek a series of filling cisterns within the animal's nervous system. But although it is only an analogy, it does help to visualise the way that so-called 'drive', or the equivalent physiological changes in an animal's internal state, may build up with non-performance of a particular behaviour and may influence behaviour in the future. It also helps, incidentally, to emphasise and crystallise some of the other characteristics of behaviour we have observed along the way.

Thus, for example:

- an animal only ever does one thing at a time. Thus the performance of one behaviour pattern stops – absolutely – the performance of alternative behaviours by some form of inhibition.
- a behaviour pattern, is only expressed in response to the correct motivational state (as a summation of stimuli from the internal environment) and the correct external stimuli.

One clear consequence of this is that

- if the correct external stimuli are present for a number of different responses, the one that will be performed will be the one with the highest motivational level or priority
- once a behaviour pattern is started, it is (usually) carried through to completion. It doesn't usually stop part-way through. While intensity of expression may vary for a number of reasons, the actual performance of the behaviour is unitary: all or nothing.
- it is the completion of that unitary behaviour pattern that turns off the response.
- performance of the vast majority of behaviours is commonly followed by a prolonged period of non-performance (i.e. performance of the behaviour alters the internal state sufficiently that a period of time will usually elapse before that behaviour may again be elicited).

If we look closely at these last two points, we may see that there are therefore, in practice, two separate mechanisms which will stop any particular response: a short-term stop after the performance of a fixed 'unit of performance' of that behaviour; a longer-term stop caused as a result of the change in internal state effected by that initial performance.

That this is correct can be demonstrated readily, using our example, well-worn by now, of the drinking dog. Interventive experiments of this type are, happily, now less-frequently performed and would require very careful justification, but a relatively commonplace surgical procedure in earlier experimentation involved insertion of a fistula into the oesophagus of laboratory animals – an open tube inserted into the throat and discharging directly to the outside – so that anything ingested did not reach the stomach. When offered water, a dog with such a fistula will drink greedily for a certain length of time and then stop. Each drinking 'bout' is completed after a set period of time, actually measured by the passage of a certain volume of water over receptors at the back of the throat. In a normal dog, that water passes to the stomach and actually triggers another monitoring mechanism, the degree of distention of the stomach wall acting as a measure of the total volume of water consumed, before it is absorbed across the wall of the stomach and intestines into the bloodstream and changes the osmotic pressure of the blood (monitored internally by sense organs in the carotid artery, page 34). In the fistula'd animal, any water consumed simply splashes onto the floor. So although each drinking 'bout' ceases after the prescribed unitary period for performance of

that behaviour of drinking, there is no subsequent feedback to say that the stomach is distended or, more importantly to say that the animal's level of dehydration has actually diminished. The carotid receptors continue to signal that the body water levels are low. In consequence, after a short period, the animal starts to drink again – and its behaviour is in fact characterised by repeated short bouts of drinking; each bout of much the same length before it stops, but rapidly repeated.

Clearly this last example shifts away from our abstract models, and we are able to talk in terms of the actual physiological mechanisms responsible for these different elements of the control of this behaviour of drinking. We may also go somewhat further in determining the physiological basis of much of what we have summarised in our hydraulic model. From the work that has been done so far, it appears that the hypothalamus is the seat not only of what we have been calling 'drive', but of motivation as a whole. The evidence for this comes from two lines of approach, sadly rather cavalier in their methodology. Indeed, in trying to discover which parts of the brain seem to be responsible for different elements of behaviour, there have always traditionally been these same two lines of approach: in the first instance, cut bits out of the brain and see which behaviour is affected; in a more sophisticated approach, insert microelectrodes into different parts of the brain and observe the effects of direct stimulation.

Both types of experiment, both stimulation by micro-electrode and oblation, confirm that the hypothalamus is the part of the brain most closely involved in what we have called motivation. From our theoretical considerations thus far, we would hope to find specific control centres for specific behaviours or at least related groups of behaviours (corresponding, if you like, to the individual, behaviour-specific cisterns of Lorenz's hydraulic model) and both excitatory and inhibitory regions in each 'behaviour centre', where nervous stimulation may either increase or decrease motivational level. While early experiments (with rather crude electrodes) were somewhat equivocal, subsequent exploration with micro-electrodes has confirmed that different regions of the hypothalamus indeed appear to be associated with different specific behaviour patterns, although these may overlap to a degree where behaviours are of similar 'style' (drinking and eating, for example); in addition, within each focal region there are indeed distinct excitatory and inhibitory areas.

Basic survival value and the 'need' to perform a certain behaviour at specified intervals, together with an increasing tendency towards expressing a particular behaviour after a given period of non-performance, are not the only factors which may influence motivational state. We have already alluded to the significant influence of hormones and hormonal changes within the body and there are many other potential influencing factors. These also work through a specific effect on the hypothalamus. Perhaps surprisingly, from the point of view of a definition of motivation as the suite of internal factors causing a change in an animal's response to constant external stimuli (page 34), even external stimuli themselves may directly influence motivational state.

For as we have already discussed, the action of external stimuli can be at two levels – either in eliciting a particular response as a releaser, or in acting in a more general way to determine a particular state of responsiveness. You will remember that we distinguished in the last chapter between two quite distinct categories of stimuli which act to release behaviour or to modify the intensity and direction of that behaviour. Releasers appear suddenly and are impermanent, while motivating stimuli are continuously present and are a part of the constant environment in which the animal finds itself. And, if you look at it that way, there is nothing wrong with the concept that general features of an animal's external (as well as internal) environment should influence its motivational state (much in the same way as hormones tend to do, in altering its general state of responsiveness in a rather general way).

The trouble is that, while this is all very well in theory, it is in practice quite hard to distinguish the one from the other. One of the most regularly cited examples of a reduction in general responsiveness brought about by a specific external stimulus returns us once again to the courtship of the three-spined stickleback. As you will remember, (page 20) the courtship culminates with the female swimming through the nest tunnel to deposit her eggs; the male follows, fertilises the eggs and then chases the female away. In experimental situations it has been found that the frequency of zig-zagging to a standard model (which always elicits some response, since the behaviour is a fixed action pattern largely under reflex control) is reduced immediately after fertilisation, and the level of aggressiveness is increased. Manipulation of the various different factors involved (such as whether or not courtship had occurred at all, whether or not fertilisation occurred, stimuli from the female, presence or absence of eggs) showed that the suppression of the probability of showing courtship behaviour to the model female was primarily a consequence of the presence of a fresh clutch of eggs in the nest (whether the male had fertilised them or not).

This example seems fairly clear, but there is often more difficulty in establishing that the effect of a given external stimulus is through an influence on motivational state rather than simply directly, as part of the releasing pattern. Perhaps the idea that external stimuli can indeed, quite regularly, affect motivational state may be strengthened if we turn to another class of stimuli, which psychologists refer to as incentive stimuli. (This idea of incentive stimulation has been 'poached' – and corrupted – by many educational psychologists who actually often use the term motivation, when they actually mean this specific instance of the effects of incentive stimulation. Thus they commonly refer to mechanisms of 'motivating' students or complain that pupils are not 'motivated' to work, when in fact they mean that the pupil has no incentive to do so).

Throughout our considerations in this chapter, you may have noticed that we have been dealing with an increasing complexity of response: our considerations of 'drive' relate to behaviours little more complex than simple reflexes or reflex chains; the effects of hormonal state involve more general changes to a state of arousal or general attention; now we move into the realms of the influence of learning.

Incentive stimulation relies on the fact that an animal has learnt to associate a particular situation or set of stimuli with a reward of some description. Thus those same stimuli in the future, with the implicit prospect of some reward, affect the animal's probability of performing that behaviour and the frequency of repetition. Thus children do indeed tend to apply themselves more seriously to their studies if they are regularly rewarded with a sweetmeat on completion of a task. And animals, too, show similar responses. In the classical experiment of requiring a laboratory rat or mouse to run through a maze, the animal is rewarded on its arrival at the home box with a small amount of food as its 'incentive' for performing the task. Increases in the quality of food offered as a reward lead to rapid increases of interest in running the maze in the first place and the speed at which they run; the higher the quality of the reward, the more enthusiastic the test subjects, and the faster they reach the home box – a clear illustration that motivation has been increased by increasing the incentive.

In this chapter we have spent some time considering what factors may contribute to something within the animal we may call its motivational state, and how that complex of factors may affect the animal's probability of responding to a given set of external stimuli with one particular behaviour rather than another. Interactions between elements of the animal's internal state and external stimuli alter the simple link between stimulus-response of pure reflex behaviour and offer a much greater flexibility in responsiveness, in the overall probability of response, and in the "choice" between alternative behaviours at any given time.

But that does not take away significantly from our ability to predict behaviour, to analyse what is going on and to predict what will happen next – our litmus test that we really do understand what is going on. For, if we have been observing our animal for a period of time, we know what behaviours it has recently been performing: we know whether it has recently fed or drunk, so we know whether it is 'hungry' or 'thirsty'; similarly we can estimate, for a host of other behaviours, the length of time since last performance and thus estimate the relative likelihood of their repeat performance in the future. From ongoing behaviour we can also assess something of the animal's hormonal state, and from careful examination of external cues may estimate various incentives. Since an animal can only alter its internal environment, its internal state, through its own behaviour, observation of that behaviour over a reasonable period of time enables us to gauge that internal state with reasonable accuracy.

With that assessment, and close observation of the external stimuli which we know the animal has the capability of perceiving, we may still have a pretty good idea of how it will respond in any given future situation. We have already progressed a long way from simple behavioural reflexes; but we have also seen that even more complex behaviours can be explained in terms of an inter-relationship between a number of external cues and the animal's internal state; yet, incredibly enough, we are still in a position to have a fair shot at predicting what behaviour we may observe in any given context. There are, though, some oddities. Even when we think we have been meticulously careful in our analysis of the situation, an animal may still sometimes surprise us by doing something completely unexpected.

6

'Unpredictable' behaviours: redirected and displacement activities

Every so often in the study of behaviour we come across behaviours that don't 'fit': behaviour which appears to be performed out of context with the environmental situation, or the animal's recent past behaviour, behaviours which are incomplete in some way, or are neither one thing nor the other. Yet even these apparently incomplete or irrelevant behaviours are not as inexplicable as they first seem. If we analyse carefully the sorts of situation in which they occur, we observe that even these apparently out-of-place behaviours appear under very specific conditions. And if we examine the circumstances closely, we may find a pattern in even *their* appearance, and reduce them too, to the level of the predictable – or at least the explicable.

Commonly, these 'unexpected' behaviours appear when the behaviour one might perhaps expect to see in a particular situation is blocked in some way. There are three ways in which this might happen. First, the animal may be highly motivated to perform some particular behaviour but unable to express the behaviour because some essential part of the external stimulus situation is missing. We have noted above that Lorenz's flushing lavatory has spring-loaded valves and that some behaviours can indeed be expressed even if the external stimulus set is incomplete, if the motivation is sufficiently intense. But 'incomplete' is not the same as 'altogether lacking' and it is rare indeed for behaviour to be performed entirely in a vacuum.

This situation – of high motivational levels for some particular behaviour but complete lack of appropriate external stimuli – is classically referred to as **frustration**. Commonly it leads to an increase in activity and what is called appetitive or searching behaviour as the animal moves around in its environment in search of the necessary stimulus, in the way that a thirsty animal in the absence of water may search until it finds it. But in the total absence of the stimulus – and unable to locate it – the animal may perform a different behaviour altogether. Many nesting birds, for example, develop a special 'brood patch' on the abdomen. In response to changes in the levels of sex hormones, associated with the breeding season, a portion of the abdomen loses its feathers and becomes highly vascularised (an increase in the density of superficial blood vessels) in order to provide warmth for brooding eggs and chicks. In conjunction with these changes, the brood patch is also very highly innervated and the sensitivity of those nerve endings themselves is increased as another effect of increased hormone levels. An incubating bird is accustomed to the feel of eggs on the brood patch. If the eggs are removed, the bird becomes 'frustrated' and may perform a variety of apparently irrelevant activities: drinking repeatedly or preening itself over and over again with unusual intensity, even though it may have preened thoroughly only a few minutes before.

The animal may have the correct motivation for a particular behaviour and the correct external stimuli may be present to permit the release and performance of that behaviour, but it may be physically prevented from acting. Thus, in a simple laboratory experiment, hungry mice may be able to see and smell food in the food hopper of their cage, but are unable to reach it (it may be too high, or they may be prevented from access to the food pellets because someone has inserted a sheet of glass between the food and the mice). This type of situation is formally what behaviourists mean by the term **thwarting**. And animals thwarted in their attempts to express a relevant behaviour may well perform something that appears entirely irrelevant in the given situation. Like our frustrated bird, these mice may drink repeatedly, or busy themselves in repeated bouts of unusually vigorous and thorough grooming.

Finally – and, in the natural world rather than that of the experimental laboratory: more commonly – 'irrelevant' behaviours may be expressed in response to situations of **motivational conflict**. If the external stimuli are present for two behaviours simultaneously, we already know that the one which will be performed will be that of the higher motivational status or priority and that it will inhibit the performance of all other behaviours until it has been completed (page 39). But what if the correct external stimuli are present for two behaviour patterns of equal motivational strength? Occasionally we may see the performance of the two behaviours together, in rapid alternation, or in some curious composite response which patchworks together compatible parts of the two. More commonly, we see the expression of some completely different behaviour altogether, apparently out of context in terms of the current situation or the animal's immediate past behaviour.

In situations such as these, where an animal is prevented in some way from performing the behaviours of highest motivational priority, we may see a number of different types of response. Many of them are relatively simply explained: as a partial expression of one of the blocked behaviours or (in conflict situations) as an interaction of elements of the two conflicting responses. Thus we may commonly observe an animal 'starting' to do the behaviour which has been frustrated or thwarted, but because it is prevented from expressing the behaviour fully, it performs the response only in a very incomplete form; such behaviours are commonly referred to by the general but very descriptive term of '**intention movements**'. Intention movements are, in effect, the result of a reduction in intensity or frequency of the 'proposed' behaviour, so that it appears in an incomplete form, usually as the initial phases of movements or movement sequences, but remaining unfinished. These movements may be repeated over and over again many times in succession.

Where blocking results because of a motivational conflict between two competing behaviours, the intention movements of two blocked behaviours may result in a rapid alternation of behaviour, as the animal starts one behaviour, then the other, in rapid alternation. Just before copulation, a male chaffinch has conflicting tendencies to approach the female and to avoid close physical contact with another individual (which may potentially be aggressive). He does not stay still at the point of balance, but alternates rapidly between approach and avoidance, sometimes edging closer, sometimes moving further away. And indeed, while now ritualised into a formal courtship display, the zig-zag 'dance' of the male three-spined stickleback (page 20) is actually a classic case of such alternation. When a gravid female enters a male's territory, he swims in a zig-zag course: one leg of this is towards the female in an element of the attack of an intruder; the other away as he attempts to lead her to his nest: a classic alternation.

Another class of behaviours we may recognise in situations of behavioural conflict or frustration is that of the **redirection** of the motor patterns appropriate to the expected behaviour to a different object. Where our hungry mouse is physically prevented from reaching the food in its food hopper, we have noted that it may regularly make visits to its water bottle. These drinking bouts are often somewhat stylised: over-vigorous, yet incomplete. The movements in drinking are almost the same as those for feeding, and in this redirection the mouse's actions are almost intention movements for feeding – but expressed towards a different object. Such redirection does not only occur in situations where a single behaviour is thwarted or frustrated in expression: it may also occur in situations of motivational conflict, where one behaviour, while successfully inhibited in full expression by the other, is nonetheless in some way dominant. Courting sticklebacks show a composite alternation of attack, retreat and attempts to solicit the female. Territorial animals on the boundary of their territory are 'torn' between attack of a rival in the next territory and flight from that same rival.

Any animal's 'confidence' in ownership of territory (and willingness to attack an intruder) is highest towards the centre of its own territory, and declines towards the boundary. Likewise, its 'nervousness' – or readiness to run away when challenged – increases as it moves further and further towards the centre of the territory of another individual (where it itself is an intruder). In encounters on the boundary between its own territory and the territory of a neighbour, the tendencies to attack and retreat are closely balanced; the animal is strongly motivated to attack its rival, but equally motivated to flee from that same rival. In such situations the tendency to attack is usually slightly higher, but not sufficient to overcome the tendency to flight: the animal directs a ferocious attack at its own mate, or an inoffensive clump of vegetation.

In such territorial encounters, male herring gulls quite commonly redirect the attack on their own mate if she happens to be in the near vicinity, or furiously peck at a clump of grass; rival roe deer males thrash viciously with their antlers at nearby trees and shrubs, rather than directing the antlers at their opponent. But it is a delicate balance and should the rival venture across the territorial boundary onto the owner's ground, the attack will be immediate and on the flesh, so to speak; it is only in the situation of precise balance of the conflicting tendencies to attack and to run away, on the territorial boundary itself, that the aggression is redirected.

Displacement activities

All these different behaviours have their origins in one or the other, or both, of the primary responses which are inhibited through thwarting, frustration or motivational conflict. All can be readily recognised as derived from them in some way, even if they may appear stylised or incomplete, or may be redirected to a different target.

However, there is one other class of behaviours characteristic of these situations which cannot be so readily explained. Every so often, an animal responds to a situation of frustration or conflict by producing, out of the blue, a different behaviour altogether, a behaviour which appears to be completely out of context with the preceding situation. Our frustrated canary, lacking the expected stimulus of eggs on the brood patch, drinks repeatedly or preens itself with enthusiasm; our thwarted mouse, able to see and smell food

but physically prevented from reaching it to feed, suddenly grooms itself repeatedly and vigorously. These actions were seen by early behaviourists as a way of doing something when, if you like to be anthropomorphic, the animal was 'embarrassed' at not being able to do what it wanted to do. As such, these apparently irrelevant activities came to be known as '**displacement activities**'.

Drinks parties are one of my personal happiest hunting grounds for good, unmistakable displacement behaviour. Remember, you are on someone else's home ground, their 'territory'. How are they going to treat you? As friend or foe? That girl over there that you quite like the look of – should you go over and try and engage her in conversation, or are you simply looking for a put down? Worse, is that chap with her perhaps not her brother after all, but someone else trying to chat her up? Some people react by turning on a dominant act, talking too loudly, laughing too long. Others become submissive, quiet and retiring. Others again, in full conflict, smoke nervously, pluck at their own clothes, nibble on canapés they do not really like: engage in true, full-blown displacement activities.

There is no question but that displacement activities are the most striking of all "irrelevant" behaviours, *and* the most apparently irrelevant. But, in practice, they have a number of clear characteristics and are actually far from irrelevant. To start with, we may note that they are not peculiar responses observed only in situations of conflict, but are actually behaviours which are quite regular parts of the animal's normal behavioural repertoire. It is no coincidence that the examples I have used have referred often to eating, drinking or grooming. These are in fact amongst the activities most regularly observed in displacement yet are also in themselves amongst the most common behaviours of the animal's normal behavioural repertoire.

It is true that when performed in contexts where they appear as displacement activities, the behaviours often appear to be more vigorous than normal and are often not properly completed: an animal engaging in displacement feeding often does not actually ingest the food, merely picks it up and mumbles at it; an animal drinking in displacement often only takes exaggerated sips and frequently returns to the water bottle over and over again. But the behaviour is not expressed in a vacuum: even in displacement, the behaviour is clearly under the control of external stimuli, which may affect which behaviour is performed as the displacement action, and also affect its intensity.

Displacement activities are in fact selected from amongst those actions most common in the animal's normal behavioural repertoire; the choice of which behaviour is performed is influenced by what external stimuli are present, just as in its normal performance; and again, just as during its normal expression, external stimuli exercise some control of the behaviour during its performance. While a chaffinch may regularly 'indulge' in a frenzied and often exaggerated bout of preening itself when in conflict, it is more likely to select that behaviour (rather than another activity) if its feathers are already ruffled or sticky. And having started, the intensity with which it preens itself is also increased – exactly as normal – if the feathers are wet or sticky. A male stickleback, 'blocked' in some way during courtship, will often fan enthusiastically – as if fanning eggs – but the intensity of that fanning is controlled closely by levels of carbon dioxide in the water, just as in its more normal performance.

And herein lies the key to what is going on. Amongst the external stimuli surrounding an animal at any time, the ones most commonly present may be those for feeding, drinking and grooming. Indeed, the very 'frustration' of not being able to perform the behaviour now being blocked, may itself increase the stimuli for grooming by causing a surge in adrenalin, accompanied by mild sweating and partial erection of fur or feathers: a common reaction to stress. At the same time, these same behaviours of feeding, drinking and grooming are those which are commonplace in the animal's normal behaviour and actually require fairly frequent repetition. In terms of Lorenz's hydraulic model, these are behaviours whose tanks have a relatively rapid fill rate: they are, in short, behaviours for which motivational levels are always fairly high. If the expression of a given behaviour is blocked (whether physically prevented from expression, prevented by lack of some essential part of the external stimulus set or blocked in conflict with another behaviour of equal motivational status), what could be more sensible than to express in its place the behaviour next highest in motivational priority, and for which the correct external stimuli are present? And since feeding, drinking and grooming are always behaviours of relatively high motivational status and those for which appropriate external stimuli are commonly present, what could be more logical than that the animal shall perform one of these behaviours, appropriately and entirely in context?

The fact that the behaviours are often more intense than in normal performance, or somewhat stylised in appearance, may simply be due to a somewhat heightened state of arousal. This is known to lead in general to more rapid switches of attention in any case, thus leading the animal to encounter or attend to a different array of environmental stimuli, causing more rapid switches in behaviour. It may indeed be simply demonstrated that behavioural frustration (in a more general sense: including thwarting or motivational conflict, as well as classic frustration) does indeed result in an increase in general arousal, recorded by EEG as increased electrical activity in the brain. Displacement activities therefore are neither irrelevant nor unpredictable, but fit closely with our developing understanding of the way in which behaviours are controlled. Since one behaviour (that with the highest motivational status) has been blocked, the animal merely responds by performing, in context, the behaviour next in importance on its motivational agenda. It is out of context only in the framework of the behavioural situation apparently leading up to it.

'Irrelevant' or incomplete behaviours and communication

One final point remains. Behaviours born of motivational conflict – in particular, intention movements, alternation, redirection of behaviour or displacement activities – are clearly, in the simplest sense, also modes of communication, however incidental. In a territorial dispute, a herring gull seeing another begin to preen itself is actually being given a clear indication that its opponent is in precise motivational conflict. If the opponent instead pecks vigorously at the ground, the first gull is provided with the different information that the other bird is still more aggressive than submissive, but marginally so, and so on.

As a result of this secondary role that they have assumed in communicating information about motivational state, specific 'conflict' behaviours are often emphasised, made even more vigorous, more incomplete, in translation to a stage where they become shaped into an explicit rather than implicit form of communication. Many ritualised threat displays or courtship displays are developed from the basic components of various displacement, intention or redirected activities shown in that particular conflict situation (aggressive dispute: and the conflicting tendencies to attack or avoid; courtship: and the opposing tendencies to court or attack an individual intruding into one's personal space). We have already revisited the zig-zag dance of the male stickleback, which is indeed a ritualised version of such ambivalent behaviour, and other deliberate display behaviours show clear evidence of such origin from initially inadvertent signals.

In an aggressive encounter between two male mallard ducks, one may suddenly break away with a submissive posture, exaggeratedly turning its head away and touching with its bill the brightly-coloured patch of feathers (speculum) on its wing. Now stylised and ritualised, now emphasised by the specific development of the coloured patch of plumage, the original root of this behaviour is nonetheless clear: it has developed from the initial displacement preening which might have been observed in a bird neatly balanced between aggression and retreat.

Learning and its effects on behaviour

One final way in which an animal's behaviour may be modified is through learning and experience. The physiological basis of learning is complex and may involve both chemical changes and the development of new connections and linkages between nerve cells in the brain.

Simple organisms with no clear centralisation of nervous function in a central ganglion or brain can nonetheless show short-term learning. Even an amoeba may learn to avoid unpleasant stimuli and we have already mentioned one form of short-term learning in earlier chapters in introducing the idea of habituation to stimuli which prove never to be relevant. In the simplest organisms, learning appears to be by chemical change or accumulation of chemical memory, rather than by specific changes in the nerve network. Indeed, one of the oft-quoted illustrations of my own youth was of the possibility, in such simple organisms, that learnt information could be directly transmitted from one animal to another. I never traced the story in its original form – and it may well have been embellished in the telling – but the essence of the tale was that planarian worms, like many other animals, may be taught simple tasks like negotiating a simple T-maze (where you simply have to learn whether to turn right or left) in response to a food reward. Once the maze had been learnt, I was told, if the trained worm was then macerated and its contents injected into a flatworm completely inexperienced in such a task, the injected worm would nonetheless show improved ability to learn the maze, superior to that which might be expected from a completely naive animal. Planarians are actually sufficiently simple in organisation that (like a number of other creatures of relatively simple structure) they possess full powers of regeneration, following accidental damage. In an extension of my apocryphal tale, I was taught that if an experienced worm, having fully learnt the maze, was cut longitudinally into two, each half would redevelop into a complete worm and each would retain precisely 50% of the learnt task! I am not too sure that I believed that part.

In more complex organisms, learning and memory appear to be a function both of chemical changes in the brain and of an ability to grow new nervous connections and links within the central nervous system. Commonly, a distinction is drawn between short-term and long-term learning. The consequences of some experience first affect a short-term storage mechanism in the central nervous system, but this short-term memory decays quite quickly with time and is easily damaged or reversed by mechanical or electrical shock. Once learning has affected the long-term storage mechanism, its effects are permanent, and decay only with disuse. It is generally accepted that this long-term mechanism cannot be targeted directly, but that experience must first effect changes in the short-term store, and it is only from here that they may become more permanently encoded.

Types of learning

In Chapter 4 we first introduced, within the context of filtering of relevant stimuli from the environment, the idea of habituation: a relatively persistent waning in responsiveness as a result of persistent, repeated stimulation, not followed up by any consequence (not 'reinforced'). This really is an extremely simple form of learning (apparent in organisms from amoeba to man) and involves at its simplest merely a decrease in intensity of frequency of response. As we have seen, it is reversible if the stimulus is withdrawn for a period and then presented afresh, or if the stimulus is accompanied by some reinforcement of its relevance.

More familiar forms of learning involve some form of 'conditioning' where, through repeated experience, an existing behavioural response may become associated with a different stimulus from that which usually elicits it (or an existing stimulus becomes associated with a novel response). In the classic conditioned reflex, if, on every occasion that the normal releaser is presented, it is presented in association with a second stimulus, the animal learns in time to associate the new stimulus with the response and will eventually respond to this 'conditioned' stimulus alone. This conditioned learning is perhaps best known from the classic experiments of the Russian psychologist Pavlov.

In a normal reflex behaviour, dogs (or any other mammal, for that matter) begin to salivate in the presence of food. Such salivation in response to the sight or smell of food is an appropriate physiological preliminary to the act of feeding, since it involves the secretion of salivary fluids which lubricate the food on its passage through the foregut and begin to digest its soluble starches. But Pavlov, in his experiments, rang a small bell each time (and every time) he brought food to his dogs. To the dogs, the ringing of the bell became an integral part of the stimulus set associated with the arrival with food; in time, they began to salivate, in a true conditioned reflex, in response to the ringing of the bell alone.

This is surely an example everyone knows, yet in truth, conditioning of this kind (in linking a novel stimulus to an existing response) is actually a relatively unimportant part of the learning process under natural, rather than experimental, conditions. More fundamental in influencing a much wider range of behaviours is the form of conditioning which links an existing stimulus with a hitherto unrelated response – in what is generally referred to as 'trial and error' learning.

A separate grade of this form of learning has become known as 'insight learning': as its name suggests, this involves the production of new adaptive responses (by quite advanced animals) as a result of insight or 'thought' processes rather than experience of physical trial and error. A captive chimpanzee faced with a banana or other food item outside its cage and beyond its reach may slide a pole underneath the bottom of the cage front to hook the fruit back into the cage where it may be consumed, even though it has never seen anyone do such a thing before. The 'solution' to the problem is novel and results in the development of a new behaviour, although that in itself may merely be a specific extension of a more general family of behaviours already present (the capacity to use tools).

All these different forms of learning intergrade and all show a basic common operator: all of them require some reinforcement for the stimulus and the response to become associated. All of them (except for habituation, which is due to lack of such reinforcement) may thus be referred to as different forms of associative learning and all take a period of time to establish. In classical learning experiments, an animal faced with a choice (a left or right turn in a simple T-maze, stepping onto white or black squares on a chequerboard) is rewarded (usually with some food reward) for making the 'correct' choice and/or punished (with a mild electric shock or other aversive stimulus) for an incorrect choice. To begin with, the animal makes an equal number of 'correct' and 'incorrect' responses, but over time, the frequency of correct decisions increases and the frequency of incorrect responses declines as the animal 'learns' the task. Eventually, when fully trained, the animal always makes the correct response when faced with the given task, and the gradual improvement in performance accompanying this learning process may be presented in graphical form as what has become known as a classic 'learning curve'. In its correct usage, the steepness of the slope of this learning curve reflects the speed at which the animal perfects the task (judged by the number of trials it takes to accomplish this). Sadly however the idea of learning curves has now been (mis-)appropriated into common parlance and is often misused in place of the actual concept of learning itself, as in, 'I've had a bit of a learning curve' or 'There's been a bit of a learning curve on this one'.

Although presented here in terms of a series of correct or incorrect responses, the same learning process, in which a successful response brings the reinforcement of some reward, is involved in any form of associative learning. There are, however, other forms of learning which seem not to have this same requirement for reinforcement in order to establish the response.

If a laboratory rat is allowed daily access to a complex maze – but one in which no food is offered in the goal box – it will spend a considerable amount of time exploring the maze; however, it may make a large number of turnings into blind alleys, and its 'success' in running the maze improves only slowly through time. Surprisingly however, if after a number of days of such preliminary exploration, the animal is given a reward upon reaching the goal box at the end of the maze, its performance improves to the same level as that shown by animals which have had the same number of experiences of the maze, but have been rewarded throughout. Clearly, the unrewarded rat has 'learnt' the maze just as well as those receiving regular reward: and in this case reinforcement appears important not to the learning process itself, but rather to the subsequent performance of that learnt behaviour. Such instances, in which an animal has clearly learnt a new behaviour successfully, but that learning is not immediately apparent in any change in its behaviour until it emerges, so to speak, fully formed, are often referred to as a form of **latent learning** – the behaviour is learnt, but remains dormant, or latent, until the experience is required on some subsequent occasion.

It is not the aim of this book to cross the borders between natural animal behaviour (ethology) and experimental psychology; indeed, in general, my recourse to laboratory experiment in these pages has been limited to the appropriation of some suitable experiment which illustrates some point of debate more clearly because of its essential simplicity. A more complete textbook of experimental psychology would at this point rehearse a further and wider array of types of learning, revealed by experimentation and through the use of such devices as the so-called Skinner box. But I have no wish to enter this more esoteric arena. Sufficient for our purposes here, in exploring the behaviour of animals around us, is to note that the different forms of associative learning, and possibly the phenomenon implicit in examples of latent learning, may come to modify behaviour patterns through an animal's own experience, resulting in greater flexibility and, perhaps, greater individual variation in behaviour, at least among more advanced organisms, in which developmental experiences and opportunities for learning may vary considerably between individuals. There is, however, one final type of learning which we should consider before we leave: the phenomenon of imprinting.

Imprinting

Imprinting is a particular form of learning with somewhat curious characteristics. To begin with, it occurs extremely early in an animal's life (somewhat blurring the distinctions between what may be innate and what may be learnt). Indeed, it can only occur over a very restricted period in early life, a narrow time window lasting only a few days at most. Yet it has very pronounced effects both upon the immediate behaviour of the imprinted animal, and upon its later behaviour as an adult.

Imprinting in the juvenile has enormous survival value, for it is the mechanism by which young animals come to recognise their parents: thus learning from whom they should try to beg food and, most importantly, who to run to for protection when threatened by predators. The phenomenon was first recognised in nidifugous birds such as ducks and geese, whose young hatch fully-feathered, and leave the nest almost as soon as they are hatched; it is characterised by the fact that, as soon as they do leave the nest, they follow the first moving object they see (usually, of course, the mother). It was recognised

because it sometimes goes wrong and chicks imprint on some other, often inappropriate object. Chicks hatched in incubators may imprint upon a bunch of feathers, or a mechanical 'hen', as long as this is moving.

Newly-hatched chicks can even be imprinted upon a moving rubber ball but the classic examples of misplaced imprinting are, of course, those in which animals are allowed to imprint upon human foster parents. The pioneering behaviourist Konrad Lorenz is typically remembered for being followed everywhere by his coterie of greylag geese.

Imprinting has a number of interesting characteristics. As we have noted, it occurs during a very limited period only, early in life. Its effects are irreversible and operate at two levels or, perhaps better: over two time scales. The immediate effects of imprinting, at least for animals which leave the nest at the time of birth or shortly afterwards, are that the juvenile animal 'learns' who is its mother and follows her closely when it leaves the nest, assuring that it does not get lost, and is always close to the mother for protection during the juvenile period when it is most vulnerable and is itself defenceless.

But imprinting is not simply restricted to such nidifugous birds: it also occurs amongst altricial species (birds born naked, who remain in the nest for a considerable period before fledging as relatively independent mini-adults) and among mammals. These birds and mammals, of course, do not necessarily need to 'learn' to follow mother. And here is where the secondary characteristics of the imprinting process emerge: the learnt impression may be generalised from the particular to the general and appears important in enabling animals to 'recognise' other members of their own species, learning what characterises them by extension, or generalisation from the characteristics of the mother. And this early learning process, operative over only a tiny period of time soon after birth, determines the entire future reproductive behaviour of the animal: from its experience of its mother in those first few critical hours, it 'learns' the characteristics of future potential mates.

This actually makes a lot of sense – at least for male offspring. Adult males of most bird or mammal species have brightly-coloured plumage or pelage whose colour pattern is species-specific, or have immediately recognisable secondary sexual characters, such as antlers in male deer. Innate recognition of potential mates, for females, may thus be relatively easily pre-programmed from birth.

By contrast, females of many bird species have dull, drab or rather nondescript plumage: deliberately unostentatious so that they are better camouflaged when incubating eggs or brooding chicks; many female mammals are also rather unremarkable. Females of many different species of duck, for example, are all rather much of a muchness: discrimination between these by unequivocal characteristics might be rather more difficult to pre-programme. But learning the characteristics of their own mother, and then generalising that 'pattern' from the particular to the general, will allow juvenile males to develop a recognition template for future mates.

That this effect on the future sexual choices of males is true can readily be demonstrated in simple experiments. The German behaviourist Klinghammer, for example, transposed clutches of eggs from incubating zebra finches and placed them in the nests of society finch females (a related species), so that each hatched and reared chicks of the 'wrong' species. The chicks were removed at fledging and housed individually until mature. Then, given a choice of sexual partners, all females preferred their own species (responding presumably to innately-determined recognition of the male's colouration and courtship display). Male zebra finches from the experimental broods, however, always attempted first to court society finch females as potential partners. In a separate experiment, Klinghammer hand-reared budgerigar chicks, which became imprinted upon his hand: begging for food on the appearance of the hand, snuggling into the hand for protection and shelter. As adults, even after two years of normal courtship and breeding, hand-reared males would at once begin to court – and attempt to copulate with – a human hand if re-exposed to that stimulus.

Given this function, or at least this consequence, of imprinting in determining the whole future reproductive behaviour (at least of males), then its other attributes are clearly extremely important: the ability to extend from the particular imprinted object to the general class of objects of that type (not actually very common in other forms of learning); the fact that imprinting can occur only very early in life and over such a restricted period (which will, in normal circumstances, ensure that the juvenile does indeed imprint upon its mother); these are no longer rather curious characteristics of the phenomenon, but essential components of a fundamentally important process.

8

Behaviour genetics and the evolution of behaviour

If behaviour is truly adaptive, then not only must it show a degree of flexibility in an animal's lifetime and an ability to change in response to experience, through learning, it must also be able to respond – at the species level – to changing circumstances and changing environmental needs, under the influence of natural selection. In short, if it is to be truly adaptive, behaviour must evolve. It is now generally accepted that evolution of behaviour in this way may occur conventionally through the accumulation of directional changes in the genes, but also through a form of 'cultural evolution' in the transfer between generations of adaptive, but learnt, responses.

Behaviour, of course, has a normal genetic basis, just like any other physiological character; or, strictly speaking, the potentiality is genetically determined, the actual expression of that behaviour (again in common with a considerable number of physiological or morphological characteristics) being greatly influenced by the environment of development, as we have seen. In fact, the influence of development is more marked on behavioural characteristics than perhaps on any other, and it is often extremely difficult to decide which aspects of a given response may be genetically encoded, modified by the environment of development (often called the ontogeny of the behaviour) or modified in later life by learning. Thus the study of the genetic basis of behaviour is really quite complex.

Further, since behaviour in general is relatively complex in structure and requires the co-operation of a number of different organ systems, it rarely has a simple or straightforward genetic basis, but is usually under the control of a number of gene systems. Simple manipulative experiments (such as those familiar ones of Mendel and his peas) are therefore rather rare. In practice, there have been three broad approaches to the study of the genetics of behaviour: the comparison of the behaviour of strains known to differ (but where the genetic basis of that difference may be unresolved), the comparison of known genotypes – animals whose precise genetic difference is established – and experiments involving artificial selection. These last provide a way of demonstrating that behaviour is indeed a heritable characteristic, and may offer some ideas as to how the particular behaviour studied is genetically coded.

In seeking to illustrate these different approaches and to establish something of the genetic basis of behaviour, I have tried very hard to find examples other than the dreaded laboratory fruit-fly so beloved by general geneticists. But it is difficult indeed: fruit-flies (*Drosophila*) have the huge advantages that firstly their entire makeup is actually controlled by a rather small (and thus manageable) number of genes, secondly that their entire genetic makeup is actually known (and all the different genes identified – something which is still relatively rare in more complex organisms). In addition, they breed rapidly in the laboratory and there are already a number of different genetic strains ('mutant strains') in culture which differ in known genetic ways. The reader must please bear with me, therefore, if at least we make a start in laboratory culture.

Selective breeding experiments

Aubrey Manning, Professor of Natural History at the University of Edinburgh, undertook a series of simple experiments in the 1960s in which he picked from laboratory populations of fruit-flies the first ten pairs to mate in a particular population and then the last ten pairs. He separated them and repeated the exercise for a number of generations. Thus from the offspring of the 'fast-mating' pairs he selected again the first ten pairs to mate when they reached adulthood; from the 'slow-mating' culture he picked again the ten slowest of their slow offspring.

Over a period of time he managed to establish two distinct fast-breeding and two distinct slow-breeding strains of fruit flies. Clearly the behaviour was genetically encoded and heritable. Crosses between the two fast types or the two slow types produced flies intermediate between them – still fast or still slow. Crosses between flies from fast × slow strains produced populations halfway between them for the speed of mating: the classic result when any particular characteristic is controlled not by a single gene but by the interaction of many genes together.

Comparison of known genotypes

Let us stick with *Drosophila* for one further example. Amongst the various mutant types created by generations of laboratory culture, there is a fly which is yellow in colour (as opposed to the normal black colouration of 'wild type' flies); this yellow colour variant is known to be the result of a mutation of one single gene. In *Drosophila*, however, individual genes seem to code for more than one characteristic and it was found that, in addition to the yellow colouration, there appeared a difference in behaviour. Yellow males take longer in their courtship of females than do wild-type flies.

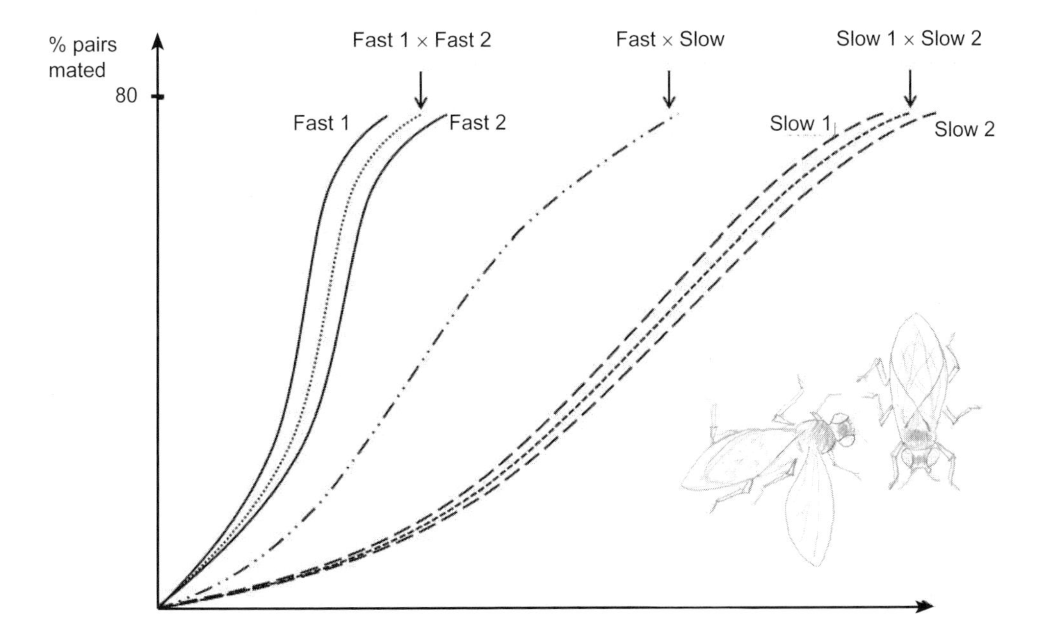

Aubrey Manning's experiments with fruit flies established fast-mating and slow-mating strains. When these strains were crossed, they produce flies intermediate between the two parents

We may dismiss the obvious possibility that this is because females simply don't like the yellow colour, disliking the colour itself, or mistrusting the fact that it is in some way 'different'. Experiments carried out in total darkness, or with females whose antennal sense organs had been removed, confirmed that yellow males still took longer to secure matings.

Courtship in these fruit flies consists of a number of separate components. A male fly approaches a female and taps with his forefeet. He then positions his body to face the female head-on, and, while standing quite still, begins to vibrate his wings at high speed. This actually produces a high-frequency humming sound. If the female still appears receptive (and has not simply left the scene) he circles around her to the rear and begins to lick her genitals. If the female accepts this, too, he proceeds to mount her and mate. The female's role in all this is more limited: in either passive (or maybe resigned) acceptance, or in active repulsion. If we examine the relative amount of time spent by males in each of these activities, we discover a significant difference between yellow males and the rest; they seem to spend significantly less time in wing-vibration, yet in studies of 'normal' flies, wing-vibration and genital licking seem to be the key behaviours in getting a female to accept their overtures. The slow mating rates of yellow flies appear to be due to a reduced tendency to vibrate, coded for by the same gene as controls the yellow colouration.

Comparisons of known strains

Let us now escape from the laboratory back into the "real world", to look at what we may learn from differences in behaviour between established strains or breeds of animals. Scott and Fuller studied the development of dominance behaviour in litters of puppies from five different breeds of dog, all raised under the same, controlled conditions. They tested dominance by offering a bone to pairs of dogs from the same litter and observing the behaviour that followed. They considered that the breed showed complete characteristics of behavioural dominance if one puppy retained the bone for eight out of ten minutes and if, when the bone was removed and given to the other puppy, the first animal promptly regained possession. If one puppy monopolised the bone for eight out of ten minutes, but when the bone was removed from it and given to the other, the first did not attempt to regain it, or failed to regain it, they considered that the breed showed partial, but incomplete dominance. No elements of dominance behaviour at all were recorded if the bone was ignored, or the two puppies shared possession.

Scott and Fuller recorded very distinct differences between the breeds, with cocker spaniels and beagles showing little or no elements of dominance while fox terriers showed consistently high levels of dominance; basenjis initially showed little expression of dominance, but responses changed over time (from 5–15 weeks of life), such that by 15 weeks they showed as complete a dominance behaviour as fox terriers. Subsequent genetic experiments suggested that these differences were controlled by two single genes: one controlling dominance, and the other a tendency to avoid confrontation.

Two fox terriers squaring up over a disputed bone

We may also consider an elegant series of experiments by Rothenbuhler, looking for the genetic basis of differences in brood cleaning behaviour in honeybees. We may clearly distinguish two separate strains of honeybees, one of which is resistant to American foulbrood (an infective fungus which rapidly kills the colony) and one which is not. The difference is due to a simple difference between the strains in behaviour. Worker bees of 'hygienic' strains, when tending the colony's brood combs, will uncap any cells found to contain an infected larva and remove the dead grub. Bees of 'unhygienic' strains do not.

If bees from unhygienic strains were crossed with bees from hygienic strains, the offspring were found all to be unhygienic – results which suggest that this might be the exception to our more general rule of simple Mendelian inheritance through a single gene: hygienic or unhygienic. More detailed analysis, however, revealed that this behaviour, too, while indeed under strict genetic control, was regulated by two sets of genes: one controlling the behaviour of uncapping the wax cells containing dead larvae, the second controlling the behaviour of removing the dead grub.

Let us take one final illustration. As we shall discuss in more detail later in this chapter, adult honeybees may communicate to others in their hive the exact location of patches of flowers with a good supply of nectar and pollen in the immediate local area – indeed up to distances of 500 metres or more. They do so by a characteristic 'dance' performed by a returning forager on the food combs of the colony's nest or hive. Karl von Frisch discovered that there are in fact three variants of this communication dance. A 'round' dance, performed by moving in alternating circles to the left and to the right, is used by foragers to indicate the presence of a nectar source close to the hive; the food patch is sufficiently close that no indication need be given of direction, since the type of flowers is clear from the scent still carried on the forager's

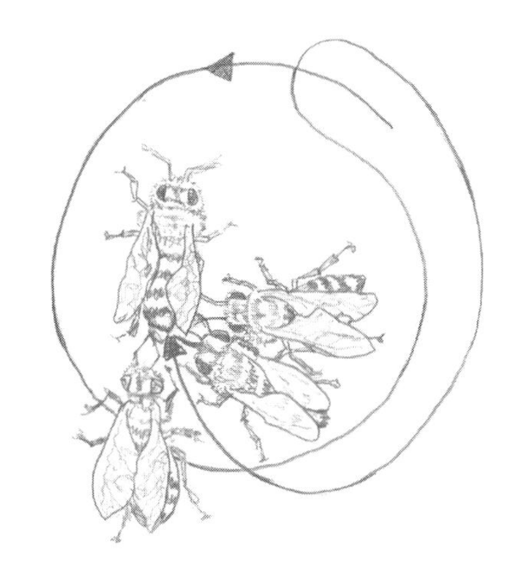

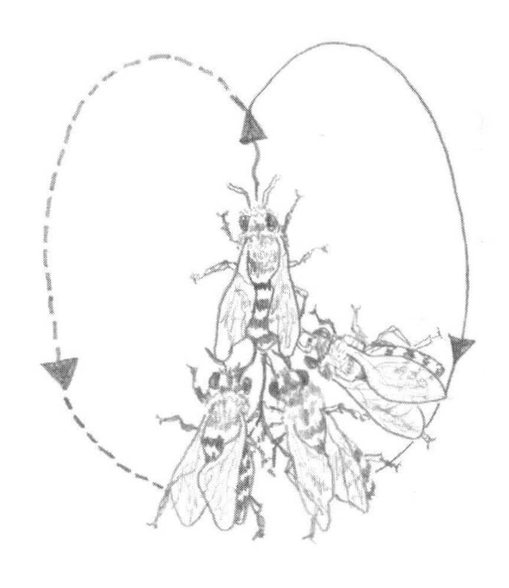

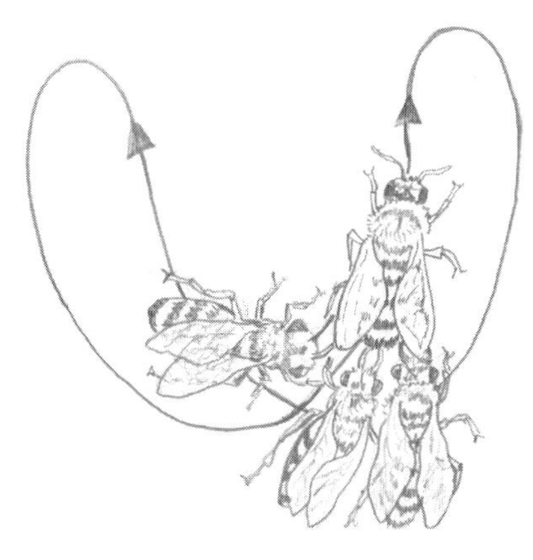

Round, waggle and sickle dances of the honeybee. Figures on this and the following page are adapted with permission from the August 1962 issue of *Scientific American*, a division of *Nature America, Inc.*

body, and other workers merely have to search for a patch of flowers of the right kind.

The 'waggle' dance indicates both the distance and the direction of a nectar source somewhat further away. The dancing bee moves in a figure of eight around a central straight line, waggling her abdomen from side to side as she progresses through the straight portion of the dance. The number of wags given during this portion of the dance indicates, to the bees clustered around her, the distance of the food source from the hive, while the angle of the dance from the vertical gives information about direction; the angle of the line from the vertical reflects exactly the angle subtended at the hive between the food source and the sun.

Finally, some bees may also perform a 'sickle-shaped ' dance, to indicate food sources at intermediate distances. The bee moves in an open figure eight, with no clear straight section; precise distance is not indicated, but the approximate direction is signalled by a straight line taken between the two arms of the sickle.

Von Frisch found that only Italian strains of honeybee actually performed this last, sickle dance, while Austrian bees danced only the round and waggle dances. Both strains used the round dance to communicate the presence of a food source close to the hive (up to three metres away), but while Austrian honeybees continue to use the same

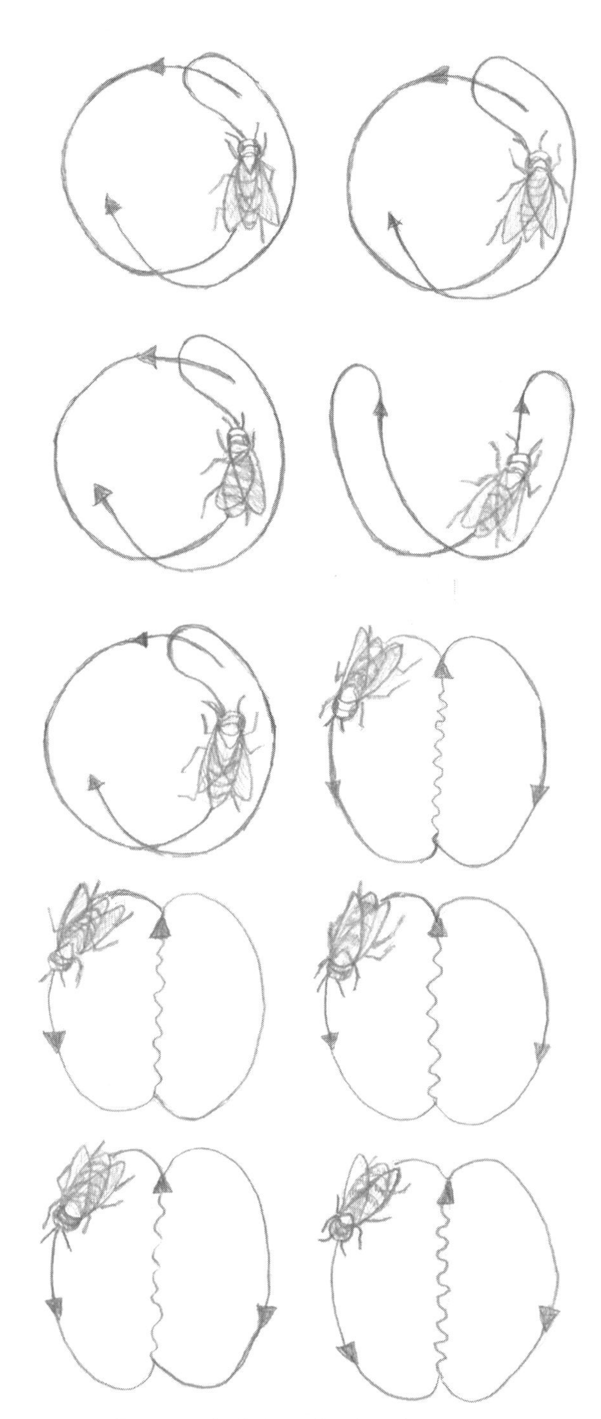

Austrian bees, on the left, and Italian bees, on the right, use the different forms of dance differently to communicate food sources at increasing distance from the hive (top to bottom). Only Italian bees offer the extra subtlety of the sickle dance (After von Frisch, 1962)

round dance for food sources up to 60 metres from the hive (before switching to the waggle dance for food patches at distances greater than this), Italian bees only use the rather uninformative round dance to tell other colony members about those food sources very close to the hive in that first three-metre band. The sickle dance is used to convey information about food sources between three and 15 metres from the hive, while the waggle dance gives full information about distance and direction for any food source at more than 15 metres' distance.

Italian strain bees also differ from Austrian bees in their colour; showing much clearer striping of black and yellow, while Austrian bees tend to be much duskier. Genetic experiments involving crosses between the two strains showed that, indeed, the characteristics of the dance were genetically encoded: some of the offspring included the sickle dance in their communication repertoire, while others did not, and those which showed the dance were those which also inherited the Italian colour pattern of black and yellow.

For those to whom genetics is a closed book or, worse, who switch off at the mention of something they expect is going to be too difficult, the take-home messages from all this are actually quite simple. These various examples and experiments simply confirm that behaviour does indeed have a genetic basis, just like any physical characteristic, and the potential for that behaviour is heritable, meaning that it may be passed from parents to offspring. In effect, it is indeed only the potential for the behaviour which may be shown to be genetically determined: as we have already discussed, its actual expression may be modified in maturation, learning – and, in day-to-day performance, by the effects of motivational state, hormones and stimulus intensity.

The evolution of behaviour

Such demonstrations do make it clear that, with a classical genetic basis, behaviour may be subject to selection and thus may evolve to improve its fit to changing circumstances. By definition, evolutionary change is a gradual process and it would be hard to demonstrate adaptive changes in one or two generations – beyond the results of purely artificial selective processes, as in Manning's fast – and slow-mating strains of fruit flies. We may sometimes, however, get a hint of the steps evolution may have taken by looking at a series of behaviour patterns demonstrated in more primitive or more advanced animals alive today.

Let us return to the communication dance used by honeybees to inform fellow members of their colony of the whereabouts of good food sources; such a sophisticated display hardly emerged fully-formed in honeybees. If we look across a range of different bee species, we discover that many other species have dances to communicate information about food sources, but that the dance takes a variety of different forms (more different from each other than the differences in 'dialect' between Italian and Austrian strains of honeybee); we may arrange these in an ascending order of sophistication, which may give some clues as to how the dance in its most advanced form may have evolved.

In relatively primitive species such as the stingless bee of Ceylon, the communication dance is hardly a refined 'dance' at all, in the sense that it is ritualised or 'coded'; rather, the returning foragers just bump into other members of the colony or jostle against them on their return to the hive, probably still carrying on their bodies the scent of the flowers from which they have been feeding. Other worker bees simply go out of the hive to search for that type of flower. Although there is some evidence that the intensity of jostling indicates the richness of the food source, no information is given about distance or direction, and we may see this as perhaps the origin of the more refined displays of other species: almost as an incidental consequence of colonial living, returning workers unconsciously convey information about where they have been successful in foraging. We may presume that, eventually, some primitive sort of dance may have developed indicating distance, perhaps similar to that of the present waggle dance – indicating distance by the number of wags in the straight section. This development is speculative, but certainly both round and waggle dances appear in the repertoire of all other colonial bee species which have been studied.

But all this while, we may presume that the dance has been horizontal and that where communication of direction has evolved, it has been communicated through the dancer taking up a position directly towards the food source. We may observe this still in the dances of another relatively primitive species, the Asian dwarf bee, which has a fully-fledged round dance and waggle dance, but which dances on the upper surface of the nest comb, positioning itself carefully at an angle to the sun identical to that between sun and food source. The dwarf bee can only dance on such a horizontal platform, and cannot dance if it cannot see the sun. We must assume that transposition of the dance to the vertical, and communication of the angle between food source and sun as that angle from the vertical of the straight portion of the dance, as in modern honeybees, was a later development required to adapt the dance to a vertical comb. We may see evidence of an intermediate step in another Asian species, the giant bee; in this species, the waggle dance has been transposed from horizontal to the vertical but, like its tiny relative the dwarf bee, the Asian giant bee still needs to see the sky and the sun if the dance is to convey any information about direction.

Honeybees are, of course, well-known to be social insects, living together in colonies of related individuals – as are many other hymenoptera, like the familiar European common wasp which plagues summer picnics. But within the wasp family as a whole, we may identify a whole range of different levels of sociality. The European beewolf of pages 11 and 12 is a good example of those species which retain a completely solitary habit. The female wasp digs a nest burrow and provisions it with a paralysed prey item on which she lays an individual egg, before simply abandoning it to its fate. But there are other solitary species which, just like the beewolf, lay their eggs in a nest burrow provisioned with paralysed prey, but then return to the nest burrow with further a prey item during larval growth; others again make regular visits to the nest throughout larval development. Yet other wasps make a single permanent nest in which all the eggs are laid, regularly making hunting trips and returning over and over again to provide their grubs with additional food. And in other species again, the young, once grown to maturity, remain with the nest and assist with the rearing of the next lot of young. In the common wasp (*Vespula*), the full colonial organisation of the most social species is seen, with reproductive suppression of daughters and the development of a full caste system of a queen, drones and workers. It

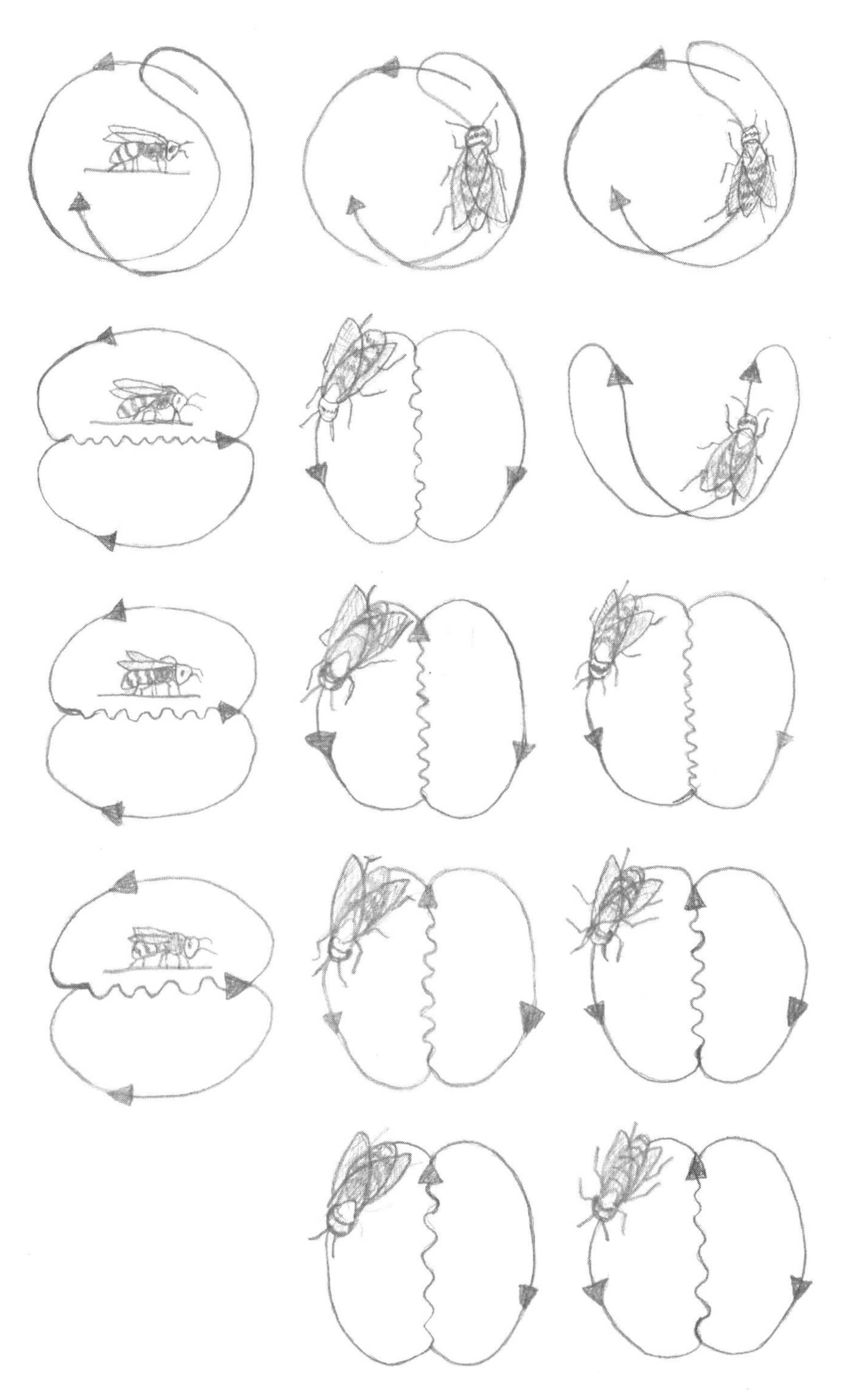

The communication dances (left to right) of the dwarf bee, the giant honeybee and the Italian honeybee; once again, the sequence of images from top to bottom illustrates alterations to the style of dance with increasing distance of the food source from the hive. Figure adapted with permission from *Scientific American* (August 1962), a division of *Nature America, Inc.*

is tempting to arrange these different species, too, into a logical sequence and suggest that these may reflect steps in the evolution of social from solitary insects.

However compelling such arguments may be, however, any suggestion that this may indeed reflect discrete steps in some evolutionary progression is purely conjectural, as it is also in the case of our earlier example of the 'evolution' of the honeybees' communication dance. The problem is that we do not know that the different 'steps' may be arranged in the convenient order in which we have placed them from the most primitive to the most 'advanced' – or even that they are true steps at all. An apparent 'series' extant today doesn't really represent an evolutionary series at all. No one 'stage' is a direct primitive precursor of the others; since all are extant to the present time, and all have had the same length of time to evolve, each must be equally well-adapted to its own local circumstances, and may not represent any effective series at all.

We must urge caution, for a similar subconscious desire for 'order' resulted in early authors arranging the remains of the foot bones of various species of fossil horses into an elegant sequence which showed not only increasing size through time, but progressive stages in the reduction of the number of the toes in the foot from five to one, and a progression from plantigrade to digitigrade stance (from walking on the spread palms – believed to be the primitive condition – to walking on the tips of the toes). Arranged in their proper sequence, the fossil remains offered almost unbelievably perfect support for the accepted wisdom of a progressive development from small, primitive plantigrade animals like *Eohippus*, to modern horses with long limb bones, with the body raised onto the tips of the toes and with the number of digits reduced to a single central toe on each foot: animals perfectly adapted to running fast over hard ground.

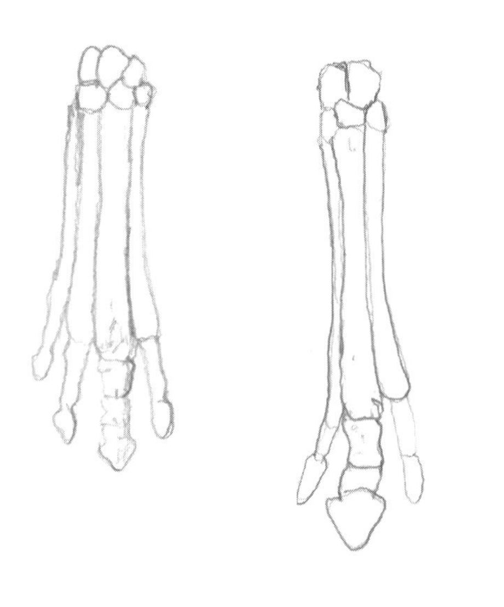

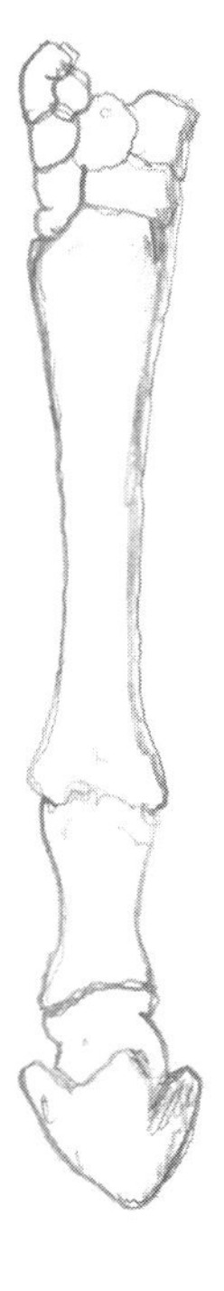

The foot bones of *Orohippus*, *Miohippus*, *Hipparion* and the modern horse, from the original drawings of Charles Marsh (1874)

Sadly, more recent re-examination of this classic fossil series has shown it to be all too convenient; in fact, the various different fossil horses of this (and subsequent) series do not represent a single progressive series through time at all, but simply reflect independent and unconnected adaptations in their own time to a series of different environments.

Perhaps the most convincing evidence one can produce in this sort of context of actual evolution is from a comparison of courtship behaviours of closely-related species. In the origin of new species, courtship behaviour and display is one of the most speedily changed of all behaviour patterns, since it is through that display that females may recognise conspecific mates, and breeding attempts are most readily restricted to other members of the 'new' species. Because of its role in ensuring reproductive isolation in this way, there is a strong selection pressure for early modification of courtship, expressed through variations on a common theme in the displays of closely-related species, with simple changes made, merely sufficient to stop the species from responding to the courtship of the other related species. Since these behaviours all stem from the same common stock (the species are indeed all very closely related) we may use them to explore evolutionary changes in that display.

One of the most pleasing examples we may find of this is shown by differences in the head-bobbing display of the group of spiny or 'fence' lizards in the genus *Sceloporus*, studied by Hunsaker in the early 1960s. Courtship in these lizards – like that in the related agamids – consists of the male facing the female with forelegs braced rigid and then bobbing the head in a series of abrupt movements. Subtle differences are apparent between species in the number and frequency of head bobs and the interval between them. The differences are small – the overall pattern of courtship has not changed, yet the differences are sufficient to identify individual species.

Similar slight variations in intensity can be observed in that section of the courtship display in dabbling ducks – the so-called grunt-whistle display – where the bill is lowered towards the water and the head then thrown back rapidly. Once again, the essential components of the display remain unchanged but different elements become more or less emphasised in different species, with some species actually flicking drops of water from the bill as the head is flicked upwards. The differences are so distinct, with more closely-related

Courtship displays of different species of ducks can be characterised by the presence or absence of a number of fundamental components. Sections of the courtship display of the mallard, gadwall and European teal are redrawn from figures by Rudolf Freund in Konrad Lorenz's article: *Evolution of Behaviour*; *Sceintific American*; December 1958

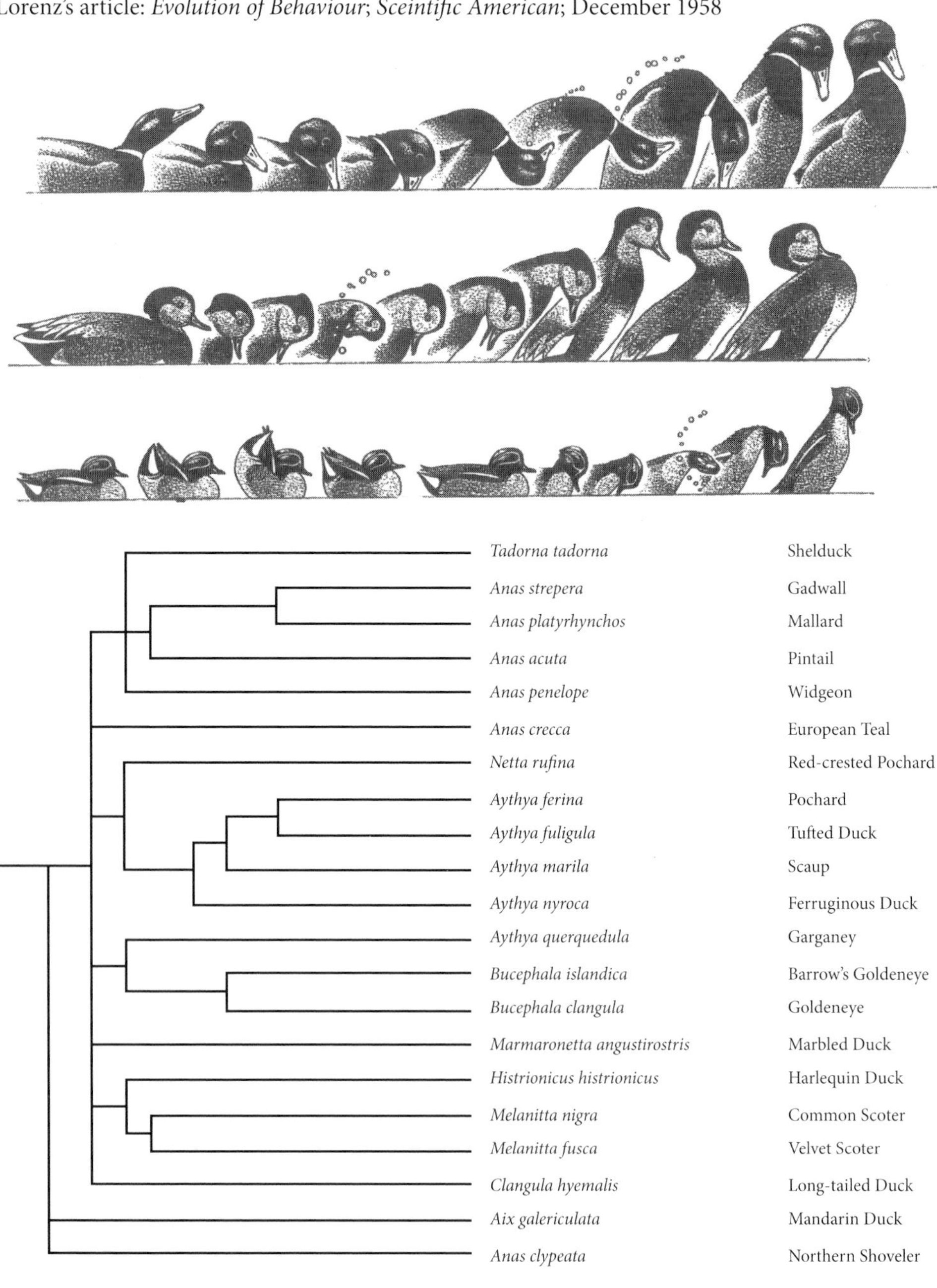

A clustering tree or cladogram of relationships between European diving and dabbling ducks based solely on similarities and differences in courtship display. This analysis (based on an earlier analysis by Konrad Lorenz) shows that even behavioural characteristics can be used to reveal evolutionary relatedness and, more importantly, shows the development of slightly different, species-specific courtship behaviours as each new species emerges. The branches in the 'tree' show how and when each species' courtship display separates from those closest to it in the diagram species showing fewer differences (and those differences being more subtle) than species more distantly related, that it is actually possible to use these displays to trace the evolutionary relationships between the different species.

The courtship displays of the different European species of diving and dabbling ducks are all based as variations on a theme; all are characterised by inclusion or absence – or relative intensity – of a number of fundamental elements: the grunt-whistle display just mentioned and other elements such as 'head-up–tail-up', standing high in the water and flapping the wings and so on. If the display of each species is characterised on simple presence/absence of the various different components and those simple descriptors alone are entered into a clustering programme on a computer (a device which looks at all the differences between each pair of species and then links the species together in clusters, in increasing order of difference), we may actually produce a cladogram of evolutionary relatedness based on the features of the courtship display alone. This shows almost exact parallel, as the clustering of species with similar Latin names clearly shows, to the pattern of relatedness we know to be true from other characteristics such as morphology or DNA. The only species which seems disproportionately different in display from its closest relatives is the Northern shoveler (*Anas clypeata*). This otherwise extraordinarily close fit between what we know to be the actual degree of relationship between the different species in evolutionary terms, and the increasing dissimilarity in display, is surely convincing evidence of a gradual evolution in displays, producing greater and greater distinction between species over evolutionary time.

The possibility of cultural evolution

The evolution of most characteristics of living organisms, whether they be morphological or physiological, must be due to the progressive accumulation of changes in genetic structure and thus may occur relatively slowly, as gene frequencies alter within a population, at least among the more sophisticated, 'advanced' animals. However, the behavioural repertoire is not simply determined by a fixed genetic basis, but may be modified by learning and by experience. Observation of another individual's behaviour may be sufficient to pass on a behavioural idea or entity, which may be further modified or 'improved upon' on transmission by some form of insight learning (page 53).

Richard Dawkins refers to these 'cultural units' as memes and if they are seen in this way as individual entities which may be transmitted from parents to children, and are subject to modification along the way, there is indeed no reason why evolution should not act on such (learnt) memes in much the same manner as on the DNA of an animal's genes. Indeed, the only difference might be that mutations of the genes are random and non-directional; natural selection 'weeds out' those changes which are disadvantageous and favours those which offer greater fitness. By contrast, most learnt behaviour patterns will be advantageous from the outset and their 'adoption' or spread within a population or a species may thus be more rapid.

There are numerous examples of this type of cultural change, with different families, or different populations of a species showing distinct and different repertoires of behaviour.

Some chimpanzee populations (*Pan troglodytes*) are exclusively vegetarian or at most eat insects and perhaps birds' eggs when they encounter them; but other populations of chimps are active hunters, killing and consuming monkeys and other mammal prey on a regular basis.

Intensive long-term studies on the Japanese macaque have also shown significant differences between the feeding habits of different bands or troops, which are of cultural origin. The monkeys are attracted to artificial foods put down for them by observers: maize, sweet potatoes and other foods. The potatoes are often caked with mud which the monkeys rub off with their hands. One day, a young female macaque was observed to dip her potato into a stream and wash it. She persisted in this habit which was copied, at first only by her own infant, but then later by other younger members of the group. Within a short time all members of the group were washing their food in this way – a behaviour quite distinct from that of neighbouring groups.

And such cultural transmission of 'ideas' or behaviours is not restricted to higher primates. We know, for example, that while the basic pattern of chaffinch song is genetically encoded, there is considerable individual variation in the final form of the song produced by an adult male. While the basic pattern is pre-programmed, we know from the work of Bill Thorpe that male chaffinches actually learn the finer details, while still juveniles, from listening to their father's singing close to the nest (a curious extension of the phenomenon of imprinting (page 54). An individual's songs are influenced primarily by those of its father, but also by the songs of other adult males singing in the immediate neighbourhood, so that, in effect, there develop distinct local variations, or dialects, in the song form characteristic of particular local areas. And we know that an individual chaffinch would pick up a different dialect if he were reared elsewhere. Indeed, one German experimenter cross-fostered a bullfinch chick so that it was reared in captivity by canaries. This bird, surrounded by other canaries, learnt their song, imitating it so exactly that it could not be distinguished from that of a real canary. Later, this bullfinch mated with a female of its own species. The two male chicks learned the canary song from their father and imitated it closely when they themselves were adults, singing like canaries rather than bullfinches.

Adult oystercatchers, while feeding mostly on crustaceans and molluscs, tend to show a high degree of specialisation in their feeding. Opening of a mussel shell for example, or feeding on crabs require very different skills, which must be learnt. Young birds inevitably learn their own mother's specialisation and thus there becomes established two quite distinct behavioural sub-populations in any group. There are even two distinctly different ways in which one may open a mussel shell: simply smashing it by hammering it on the rocks, or by inserting the beak between the two halves of the shell in a stabbing motion to sever the muscle which holds the two sides of the shell together. Different 'cultures' persist even amongst mussel-feeders.

Finally, perhaps the most widely-reported example of such cultural 'evolution' was the spread throughout the UK of a novel habit among various species of tit (mostly great tits and blue tits) in the ability to break open the tinfoil covering of old-fashioned glass milk bottles and drink the milk inside. Perhaps the development of the habit is not a great surprise. As anyone who has watched them on a garden bird table is aware, tits frequently hammer at nuts and seeds with their beaks in an attempt to break them open, and often try out novel objects by hammering them in this way. At the same time, in cold winters, when milk freezes on the doorstep, the milk in rigid glass bottles could only expand upwards, and often used to push the foil caps upwards in a dome or even lift it off the bottle top altogether; hungry tits exploring their environment for food might commonly have encountered such frozen bottles with lifted tops and discovered that there was food to be had by pecking at the milk inside. But what was remarkable was the speed with which the habit appeared and spread through the country. Further, its appearance was not in several different places simultaneously, as if in a series of independent discoveries, but seemed rather to show a pattern of dissemination like ripples from a pebble in a pond, as the behaviour was learnt and copied by a wider and wider circle of adjacent populations.

I have to confess that I have always been less than persuaded by this final example, much though I might wish to believe in it! Tits are inquisitive and innovative; and as noted above, perhaps the discovery of a novel source of food and moisture in the top-of-the-milk of glass bottles is not so unexpected. I remain somewhat dubious that the widely-reported spread of the phenomenon from the early 1920s until the late 1940s may have had more to do with changing human cultural practice and in the wider distribution of foil-topped milk-bottles than it had to do with cultural 'evolution' amongst the tits – but whether this particular example stands close scrutiny or not, the potential for a spread of learnt behaviour within and between animal populations as a type of cultural evolution itself remains beyond doubt.

9

Adaptiveness of behaviour: optimising the returns

In the first part of this book we have been looking at behaviour and asking the question 'How?' – how does an animal produce a particular behaviour in response to a given perception of its environment? But we have also established that behaviour, like any other animal characteristic, has a genetic basis and is subject to evolutionary pressures. If it may evolve, then, through the process of natural selection, it should become more and more closely adapted to achieving its purpose as effectively as possible. Yet while we believe that any pattern of behaviour must be adaptive and is shaped by natural selection to be optimally suited to its function, it is often very hard to identify the advantages that it confers, or why evolution has shaped the behaviour in quite the form that it has. In this second part of the book we are going to take this further and explore something of the function of behaviours and how each behaviour is adapted to that function, investigating more of the 'Why?' of behaviour. Why does the animal perform that particular behaviour in that particular way, rather than a similar behaviour? What is the function of that behaviour and how is it adapted to fulfil that function? It might seem that such analysis must often be purely speculative and somewhat subjective, but in practice there are a number of ways in which we can explore these sorts of questions a little more objectively by experimentation.

One of the great classics in the study of behavioural adaptation was the investigation by Niko Tinbergen and his students of why adult black-headed gulls remove broken egg shells from the nest soon after the chicks have hatched, and carefully dispose of them some distance away. Why should the parent birds do this (for they must leave the nest in order to remove the pieces of shell, at a time when the chicks are extremely vulnerable to predation, from foxes – or indeed from neighbouring gulls)? To be worth such risk, the behaviour must itself be of tremendous advantage. Tinbergen considered the possibilities:

Perhaps the sharp edge of the egg-shell might harm the newly hatched chick? – but no, it is far too thin. Perhaps the sharp edges might irritate the brood patch of the incubating parent? But again, the egg-shell is thin and easily crushed by the parent bird's weight and, in any case, minor irritation is a poor justification for a behaviour which in itself exposes the chicks to high risk. Is there a risk of disease from bacteria developing on the fragments of yolk adhering to the old shell? Unlikely and again this is surely not a high enough risk to chance loss of the chicks through predation. Tinbergen considered the possibility that the empty shell might slip over the end of an unhatched egg, preventing the second chick from hatching successfully – but no: experiment quickly demonstrates that a healthy chick is perfectly capable of chipping its way through the double layer. Amongst breeding colonies of gulls the greatest risk of all, it appears, is for the chicks to be taken by passing predators: so is this behaviour in itself a means of reducing the risk of predation in some way, despite the temporary exposure of the chicks during the actual act of removing the shell? Does the white inside of a hatched egg shell break the camouflage of the nest and thus increase risk of detection? Circumstantial support for such a notion might be drawn from the fact that this behaviour of egg-shell removal is not shown by species of gull whose eggs are not camouflaged in the first place.

To test his theory, Tinbergen conducted a number of experiments with eggs from deserted nests and showed very clearly that the risk of an unhatched egg being taken by a predator decreased directly with the distance from the nest of a broken egg shell. Thus it would seem that removal of the broken egg shell by the parent gull directly after hatching indeed reduces risk of detection of the nest and thus the risk of predation of the remaining unhatched eggs and chicks. This illustration is such a classic that it is perhaps overexposed: almost every textbook produces the same example and its repetition here perhaps lacks originality. But it is a classic with very good reason: it shows so simply and so elegantly how one may indeed set about discovering, through logical deduction and appropriate experimentation, what is at least the probable evolutionary function of a given behaviour pattern.

The function of other behaviours may be more immediately apparent. A foraging animal is clearly seeking food, because it must eat to survive. But the same style of logical approach may explain why it searches as it does: why does it select some food items and yet reject others, apparently equally palatable? Why does it abandon one patch of food for another, even when there is food still remaining in the first patch and the move involves searching for a new one? What determines when it should return to a place it has searched before? While we may not answer these questions with quite the same directness as Tinbergen explored the reasons for egg shell removal by black-headed gulls, the same principles apply. If we come up with a number of potential theories or hypotheses for how an animal forages, we may be able to test each candidate theory by experiment to see if the predictions fit the facts.

Optimal foraging

This approach to understanding the decisions animals make while searching for food is often called 'optimal foraging theory'. This is, in some ways, unfortunate terminology, for its implication is that animals necessarily do forage in an optimal way – which they may not. While evolutionary pressures act upon animals by competition with others to ensure that those with more efficient strategies survive better and leave more offspring to succeeding generations than their peers, such evolutionary change does not necessarily imply a perfect fit of behaviour to function. Simply it is enough to be better than your neighbours; as long as you are better, then there is no further selection pressure to improve. Further, while evolution does hone and refine adaptations, it takes some time to do so, and can work only over the timespan of a number of generations, yet environmental conditions rarely remain constant; they too may be changing and while adaptation in behaviour may track these changes, it may never quite catch up.

Thus the use of the term 'optimal foraging' does not imply that animals forage optimally; rather it is merely a short-hand description for an analytical device to work out why animals make the foraging decisions they do, by comparing actual observations with predictions as to the way the animals should forage if they were trying, for example, to maximise energy intake per unit time.

In practice there are a number of distinctly different ways in which animals might structure their foraging. They might seek simply to maximise energy intake - or intake of some other key nutrient (it might be protein). Alternatively, they might seek to maximise the rate of intake of food (as energy intake per unit time), or to minimise overall the amount of time spent foraging altogether (if, for example, when foraging they are more exposed to predation risk). And in effect, each possible basis for foraging actually gives rather different expectations about the way they will behave in practice. If we observe carefully what they do in laboratory experiments or in field observations, we may be able to determine what it is in fact that they are trying to optimise.

Prey selection

Any prey item consumed by a predator has a benefit in terms of its net food value, but also has a cost in terms of the time and energy taken to find, subdue and eat that food item. The balance of profitability of some potential prey items (in terms of the actual energy received compared to energy required in handling that food item in the first place) may be such that it should be ignored even when encountered. Other food items have greater or lesser profitability.

If any animal is basing its foraging decisions on either net energy intake overall, or net energy intake per unit time, we should predict that it will ignore unprofitable items, will eat marginally profitable items when it encounters them and will prefer (in absolute terms) the most profitable items. Preference thresholds will vary depending on whether the animal is attempting to optimise absolute intake of energy, or some other nutrient, or rate of intake.

In addition we would expect that (whatever commodity the forager is seeking to maximise) the relative profitability of items will also depend on their relative frequency within the environment (and thus the predator's probability of encounter with each type of food item). If the most preferred food items are present in the environment at high density, so that encounter rates are high, then the predator should concentrate on those, and ignore less profitable items even if it encounters them. By contrast, if preferred prey are rare (and thus search times are high and encounter rates low) the predator should alter its behaviour and eat any food item it encounters, whatever its profitability, as long as that profitability exceeds zero. Such conclusion results simply from the fact that energy costs involved in search are themselves significant, and thus searching longer for more productive items reduces their net profitability back closer to a level with that achieved from initially less profitable, but readily-available items. The predator will achieve overall a larger net energy intake for a limited foraging time, or a greater net rate of intake, if it now focuses equally on poorer items as well as consuming preferred items when they are encountered.

Do foraging predators do what we expect? One of the simplest laboratory studies of this type of decision-making was undertaken in the mid 1970s by John Krebs, using caged great tits as predators and presenting them with large or small pieces of mealworm as prey. (Mealworms are the larvae of a tenebrionid beetle, commonly used as a food source for captive insectivorous birds and mammals). Krebs and his colleagues estimated the profitability of the two mealworm sizes in terms of weight/ handling time and then varied the relative frequency of the two in the tits' foraging arena. The results were exactly what we might predict: when large and small prey items were present in the environment at low density, the birds were completely unselective but as the frequency of larger items was increased to a level where theoretically the birds could now do better by ignoring the small prey, the tits switched as predicted and became highly selective. When frequency of small prey was also increased so that they, too, became common in the environment, the great tits still maintained their selective foraging habits and essentially ignored the abundant, but small, prey items.

The common shrew has an extremely high metabolic rate, and most people know that these animals must consume something like their own body weight each day. With such a high requirement, one must presume that they, too, attempt to optimise the balance between energy expenditure in foraging and energy gained. In the wild they feed extensively on earthworms and other invertebrate prey; in the laboratory they, like great tits, will happily take mealworms or blowfly pupae (casters); Assuming similar handling costs, the net energy return from casters is significantly higher than that from mealworms as prey. Armed with this knowledge and in a study similar to that of Krebs with his tits, Chris Packham established that where both prey types were presented in equal numbers and at relatively high density, shrews consumed only blowfly casters; indeed, even where casters represented only some 20% of the food items provided, the shrews preyed upon them preferentially (with casters making up between 60% and 80% of all prey items consumed). Only where frequency of casters in the experimental arena was extremely low did foraging shrews increase the numbers of mealworms taken.

Such laboratory studies have since been supported with field data. Perhaps one of the best examples we may present here comes from the work of John Goss-Custard on European redshank feeding on polychaete worms (*Nereis* and *Nephthys*) on exposed mudflats. As with the previous studies, Goss-Custard measured the profitability and availability of different sizes of prey. By comparing the rate of feeding by the redshank on large and small worms at different sites (where prey availability differed), he was able to show that the largest and most profitable prey were eaten in direct proportion to their own density wherever they occurred, while the smaller (less profitable) worms were consumed not in proportion to their own density, but in inverse proportion to the density of the larger prey items, taken in fact in increasing numbers, just as we would predict, where density of larger prey was lowest.

Moving between food patches

Foraging predators often hunt for food which is clumped in its distribution and thus occurs in discrete patches in the environment. Similar considerations should affect the decision to move between patches, the predator not bothering to exploit a relatively unproductive patch at all, when encountered, but rather continuing to search for a better and more productive patch, or abandoning a patch after depleting its resources to move off again in search of a fresh one. Once again, the decision should depend on the time taken to find each item within an existing patch against the time which would need to be taken to find a new and richer patch within which intake rates would be increased. Such 'decisions' actually require some prior knowledge of the relative frequency of food patches of different quality within the environment, as well as the average search time required for securing each food item within a patch – and thus presume some time and energy devoted in the first instance to preliminary investigation and familiarisation with the local landscape. But animals do invest in such reconnaissance, and once they have that necessary information about the frequency within their local environment of food patches of different richness or quality, they may completely ignore patches of low quality when encountered, to continue the search for a patch of higher quality which will offer greater returns, as long as these are relatively frequent in the environment and thus likely to be found relatively quickly after by-passing a poorer patch. Similarly, an animal may stop feeding in a patch after it has depleted the food resources contained to a level where, while they are by no means completely exhausted, the time taken to find each new item is now increased to a point where it would be more efficient to move on and find a new and fresher patch within which the search time between individual items is significantly reduced.

Once again, the 'break point' for ignoring a poor patch in favour of continuing the search for a more productive patch will depend on the relative frequency of productive patches and those of lesser quality. And in the same way, the length of time an individual predator stays within a patch will depend on patch frequency, and thus the probability of finding a new patch where intake rates would rise above their current levels in the patch presently being depleted, and the time and energy which would be involved in locating that fresh patch.

From theoretical treatments of this 'problem', we may predict that the optimal predator should stay in any patch until its rate of intake drops to a level equal to the average rate of intake for the habitat as a whole. After a period of foraging, all food patches should be reduced to the same marginal value and this should equal the average rate of intake for the habitat as a whole. Once again, great tits foraging in the laboratory may provide our first test. Richard Cowie studied captive tits foraging in a large indoor aviary for small pieces of mealworm concealed in plastic cups of sawdust on the branches of artificial trees; cups contained different numbers of mealworm pieces, so that patches were of different quality. Tits moved between cups just as we would predict, exploiting each until the number of mealworm pieces left dropped to the average for the environment as a whole, before leaving to search for a cup which provided above-average foraging rewards. Cowie manipulated the energy and time involved in 'locating' new food patches by fitting each cup with a loose-fitting or tight-fitting cardboard lid which was easier or harder for the birds to remove. Once again, as we might predict, the birds spent longer foraging in each 'patch' when the time and energy costs of 'opening' a new patch were increased.

Return times

In such a closed system, of course (unless the plastic cups are regularly replenished), the tits will return to each starting cup in turn as soon as other patches have been depleted below the remaining reward level of that first cup. Where mealworm supplies may be replenished, they should invest some time and effort in exploration, re-checking the reward rates of earlier patches. Less work has been done on such questions in the 'real world' but some very elegant studies by the English behaviourist Nick Davies considered the return times by pied wagtails to patches of river bank, which they had previously exploited in a search for insect prey cast up by the lapping of the water on the shore, and found that the time of return coincided almost exactly with that time at which prey densities had been returned to their previous levels.

In another example, Alan Kamil studied the pattern of visits by a Hawaiian honeycreeper (or amakihi) to flowers within its territory. By numbering individual clusters of blossoms in each of five territories, Kamil was able to record the pattern of visits to flowers by individual honeycreepers. He found that visits by territory holders to the same blossoms tended to be very well-spaced in time. Intruders from neighbouring territories were far more likely than residents to visit a cluster shortly after it had been depleted (because they had no knowledge of that prior depletion). The success of the strategy is measured in the fact that intruders gain, from each blossom visited, only about two-thirds of the amount of nectar gained by a resident.

Optimising other behaviours

This approach to understanding the utility of different behaviours is not restricted in its application to questions of foraging behaviour. As a somewhat different example we may consider the mating strategy of the yellow dung fly. "Studying flies mating on cowpats may seem a rather extreme perversion", but Geoff Parker, the author of these words against himself, has raised *Scatophaga stercoraria* to a curious sort of fame within this same context of analysing the adaptiveness of behaviour and identifying what may be the optimal strategy in different circumstances.

Adult dung flies are rapidly attracted to fresh droppings. Males arrive first and since the sex ratio is something like 4–5 males to each female, there is intense competition for mates. As each female reaches a dung pat, she is leapt upon by a male who 'captures' her and mates immediately; but other males jostle the pair and attempt to displace the copulating male; not infrequently, one will succeed in kicking a rival off the female and taking over. Unlike many other insect species, female dung flies frequently mate many times: each successive mating displaces most of the sperm stored by the female from previous matings, and the last male to mate therefore predominates in the fertilisation of a given batch of eggs. Copulation can occur either on the dung itself or in the surrounding vegetation. After copulation, unless displaced, a male does not dismount immediately. He withdraws from genital contact but remains mounted in a passive phase while the female lays her eggs. Soon after the last egg is laid, the female sways from side to side several times. On this signal, the male dismounts and returns to search for new females.

Within this system, Parker identified two problems for male dung flies. First: how should they divide their time between the dung and the surrounding vegetation in order to maximise their chances of mating? And second: how long should a male stay with any one female (who may already have been mated, and who may then mate again, displacing his own sperm) to maximise his own reproductive success?

With regard to the first of these problems, habitats are seldom, if ever, uniform in their value and the distribution of receptive females within or between habitats will always be non-random. In areas of high female density, the chances of encountering a receptive female is high, but so is competition from other males; in areas of low density, lower competition may compensate for lower availability of females. Distribution of males should thus reflect distribution of females. Good habitats will be filled first, but as competition increases, so the return to be gained from poorer habitats but with lower competition may actually end up higher than that from areas which were initially better. When this happens, the 'optimal' fly should move to the poorer habitat. Ultimately – in theory! – all habitats should be packed in such a way that all males receive exactly equal mating rates. From the point of view of the individual males, their time should be distributed between the different parts of a dung pile, or between dung piles and surrounding vegetation, to achieve this distribution. In practice, as the figure below shows, dungflies are well-adapted.

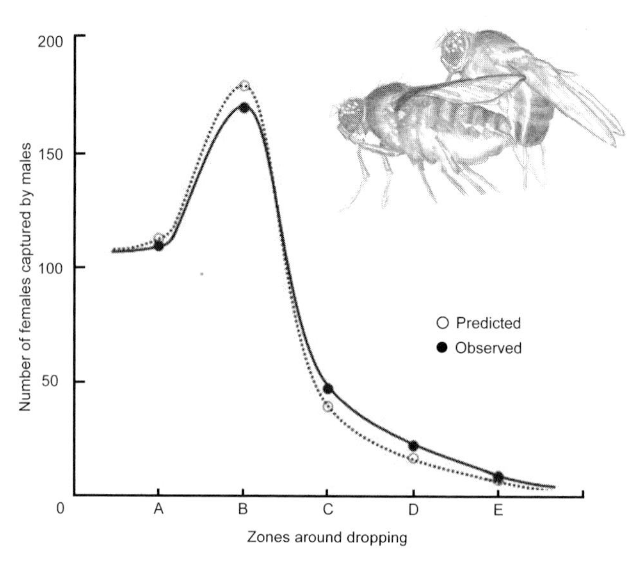

Comparison of observed and predicted number of female captures by male dung flies in a series of zones on and around fresh cattle droppings. Zone A = the dung surface; Zone B = the peripheral area of grass to a distance of 20cm from the dropping; Zone C = the band 20–40 cm from the dropping, up to Zone E, the band 60–80 cm from the edge of the dropping. (Redrawn from Parker, G.A. 1974) *Evolution* 28, 93–108)

Our second question related to the period of time each male fly should spend with any given female to maximise his overall fertilisation rate. The longer a male spends copulating with a non-virgin female, the more sperm from previous matings he will displace and the more eggs he may fertilise himself. There is, however, a cost associated with prolonged copulation, in that the more time a male spends with each individual female, the more opportunities he may miss for mating with other females. Parker carried out some very elegant experiments on sperm competition in which he interrupted a male's copulation after different time periods; this shows that the longer a second male mates, the more eggs he does indeed fertilise – but that the returns for extra copulation time diminish rapidly beyond a certain point. After a male has copulated for long enough to fertilise about 80% of the eggs, the return for further copulation is rather small and the male might do better by searching elsewhere for a new mate.

Now, consider a male which has just finished copulating with a female. Before he can copulate with a new female, he must first guard his present mate until she has laid all her eggs, or his own sperm may be displaced by subsequent copulations with other males. He must then fly off and search for a new female to court. All this takes, on average, 156 minutes. Once he has found a new female, the proportion of eggs fertilised as a function of copulation time conforms to the results of Parker's sperm competition experiments summarised in our Figure. The male cannot vary guarding time, search time or the fertilisation curve; what he can do, however, is choose a copulation time which maximises not the total proportion of eggs he may fertilise with any one female, but maximises overall the number of eggs fertilised per minute.

The optimum solution may be calculated by drawing a line from point A in the figure at a tangent to the fertilisation curve. The triangle created by this line has time as its base and the proportion of eggs fertilised as its vertical side. The slope of the hypotenuse gives the maximum possible proportion of female eggs it may fertilise per unit time. The predicted optimum copulation time is 41 minutes, not so very far from the average copulation time of a male dungfly in the field, which is in fact 36 minutes.

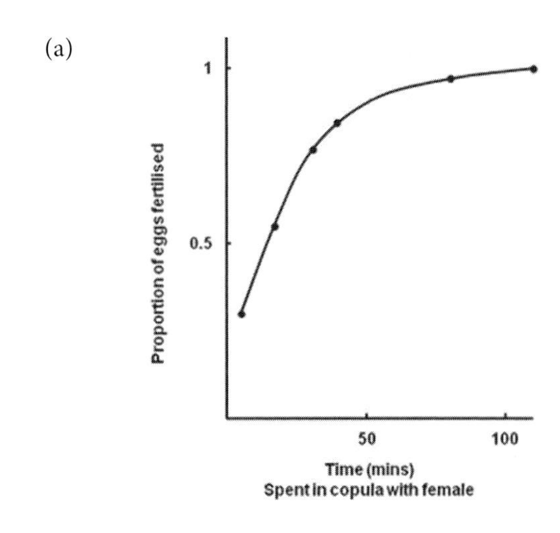

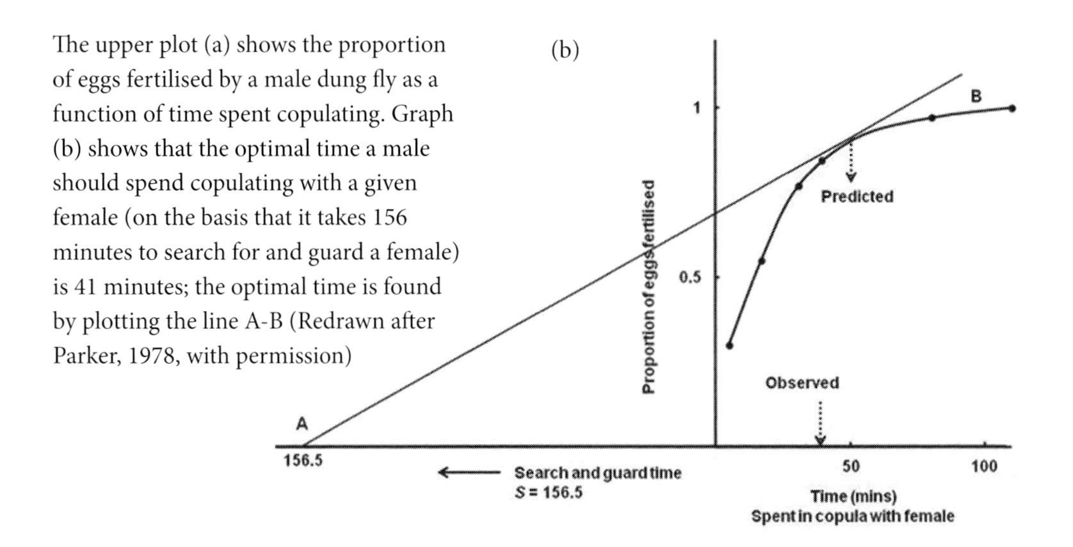

The upper plot (a) shows the proportion of eggs fertilised by a male dung fly as a function of time spent copulating. Graph (b) shows that the optimal time a male should spend copulating with a given female (on the basis that it takes 156 minutes to search for and guard a female) is 41 minutes; the optimal time is found by plotting the line A-B (Redrawn after Parker, 1978, with permission)

We may adopt a similar approach in the analysis of many behavioural choices which animals must make. Exploration of the costs and benefits of alternative behaviours which might be expressed in a given situation helps us understand why animals make the behavioural 'decisions' that they do and 'choose' the particular behavioural strategy that they do from among a host of apparently appropriate alternatives. But, of course, selection of an 'optimal' course of action does not necessarily imply a conscious or cognitive decision or that the animal 'weighs up' the costs and benefits of a number of behavioural alternatives every time it is faced with that same situation; in some cases they may actually do so, but in others the 'choice' of what is the correct behaviour to perform in those circumstances may have been established through evolution as that behaviour which, on average, delivers maximum benefits.

Further, as we have already noted, we should not assume that all behaviours are necessarily optimal. While evolution does hone and refine adaptations, it takes some time to do so, and can work only over the timespan of a number of generations, if over that same period, environmental conditions are also changing, as we considered on page 75, then while adaptations in behaviour may track these changes, they may never quite catch up to the optimal solution.

Satisficing

In addition to all this, as we have also discussed, evolutionary pressures act upon animals through competition with others to ensure that those with more efficient strategies survive better and leave more offspring to succeeding generations than their peers. As a consequence, such evolutionary change does not necessarily imply a perfect fit of behaviour to function; simply it is enough to be better than your neighbours. Indeed that may at times be an active choice in its own right – not to optimise behaviour but simply to do 'well enough'.

Satisficing, as this is called, may prove a better option than optimisation in situations where there are potentially many possible alternative options which cannot effectively be fully-evaluated or where computing the optimum solution might take too much time or energy. In this approach, one sets lower bounds for the particular objective that, if achieved, will indeed be 'good enough' and then seeks a solution that will exceed these bounds. The satisficer's philosophy is that in real-world problems there are too many uncertainties and conflicts in values for there to be any hope of obtaining a true optimization and that it is far more sensible to set out to do 'well enough' (but better than has been done previously).

Optimising or satisficing within society

Many of the examples we have considered so far are related to behavioural decisions an animal must make, but can take largely in isolation. However, as one final twist, we must recognise that animals may sometimes appear to be acting in a way that seems suboptimal when considered purely from the point of view of what might be considered ideal for that individual in isolation – but may after all be optimal if one considers what other individuals in its environment, or in its social group, may be doing. Thus, in determining an appropriate or optimal behaviour, the individual must also take into account the actions of those around it.

An evolutionarily stable strategy (ESS) is, as defined by John Maynard Smith in the early 1980s, a strategy which, if it is adopted by the majority within a population, cannot be replaced or bettered by any alternative, mutant strategy. Such ESS are by definition not necessarily strategies which would appear to confer maximum fitness to any individual, considered in isolation; however, natural selection favours traits which are optimal in the full environmental context, which in this case should be seen to include other individuals within the same population or social group. In such analysis, such strategies do deliver optimal fitness for the individual when considered in the context of the likely behaviour of others. This is an idea to which we will return, as in the next few chapters we turn, increasingly, to consider the social behaviours of animals.

In fact, more recent work has suggested that evolutionarily stable strategies (ESSs) are not necessarily, by default, behaviourally stable. Because behavioural instability implies that expressed levels of behaviours deviate from the ESS, behavioural stability is required for strict evolutionary stability in repeated behavioural interactions and we may define a separate class of behaviours which are Behaviourally Stable Strategies (or BSS). For true evolutionary stability, and to become 'fixed' within a population, a strategy must be both BSS and ESS.

Social organisation and social behaviour

As we have just reviewed, few animals, if any, exist in their world in glorious isolation, interacting with their purely physical surroundings in total independence; this world is shared with others of the same or of different species with which each individual must interact – if only, at the simplest level, in seeking a partner with which to mate. In this chapter we will review the various different levels of sociality we may encounter among animals and consider under which circumstances it is better to be solitary or fully social.

At the most fundamental level, as suggested above, animals must at least get together to mate. Natural selection acts on the success of each individual in leaving surviving offspring to breed in future generations; perpetuation of the species, and its evolution, require that a primary pressure on any individual is to reproduce. This requires some degree of co-ordination of behaviour. Even species such as oysters, which reproduce simply by shedding their eggs and sperm into the surrounding seawater, must synchronise that activity, so that the sperm and eggs are shed into the water together. In higher orders of organisation, where fertilisation of eggs is internal, male and female must not only be in the same place at the same time and in the same physiological state, but they must also develop behaviours which permit copulation, overcoming, however briefly, the natural antipathy for bodily contact and the normal inter-individual aggression.

In many cases this association is a temporary one only, sufficient only for mating, before each individual resumes its solitary habit; in other species, however, the association is of longer duration, enduring through a more prolonged breeding season, while a male perhaps protects a female and her nest and eggs, and both provide some level of parental care to developing youngsters. The length of time that young are dependent on adults for protection and support varies from species to species, and in some cases, young may remain with their parents for quite extended periods, forming stable and persistent family units. And all these levels of association are in essence designed simply towards procreation.

In a number of species, sociality may be less closely related purely to the successful production of offspring. In some cases, offspring may remain with their parents even after themselves reaching maturity, forming large familial groups, as do meerkats and their relatives, African hunting dogs or the strongly matriarchal societies of elephants, where female offspring remain, even as adults, within the mother's herd, such that a group develops of matriarch, daughters, aunts and sisters. Here, group-living affects behaviour in all aspects of life, from foraging to protection from predators, and sometimes even in terms of communality or co-operation in breeding. Such groupings truly act as a social entity, a society.

But other animals, too, may live in large social groups of **un**related individuals – like herds of wildebeest or gazelle – although here the social group is somewhat more diffuse and bonds are looser than those between groups of actual relatives. So what do these species gain from associating together?

The potential advantages of living in groups

Sociality between two individual animals in order to reproduce successfully is understandable enough, as a biological necessity, but why do some animals form these more permanent associations? What are the advantages of living in larger social groups in this way? And why do some species live socially, while others find it better to be solitary?

While we must keep at the back of our minds that not all animals do 'choose' to live in groups, let us first explore what may be the potential advantages which could be gained from a more social habit. In effect, these fall into three main areas: protection against predators, increased efficiency or competitiveness in finding food, and social advantages, such as ready access to mates, the presence of peers during development and, in the limit, actual co-operative breeding.

Defence against predators

There are a number of strategies which may be adopted by animals to protect themselves from the risk of predation. One possible method is indeed to reduce group size in an effort to reduce the probability of detection: to become solitary and relatively secretive. This approach is particularly effective for small animals and especially those who live in dense cover, where concealment is a real option (and we will find, as we explore this further, that evolutionary 'decisions' about group size do relate strongly to the sort of environment in which the animal may live). But in other circumstances, where perhaps there is little chance of escaping detection in the first place, there may be considerable safety in numbers – both in terms of getting advance warning of possible danger and in warding off attack.

When animals live together in groups (whether these are closely-knit, family groups or looser assemblages) there are simply more eyes and ears alert to possible danger, and

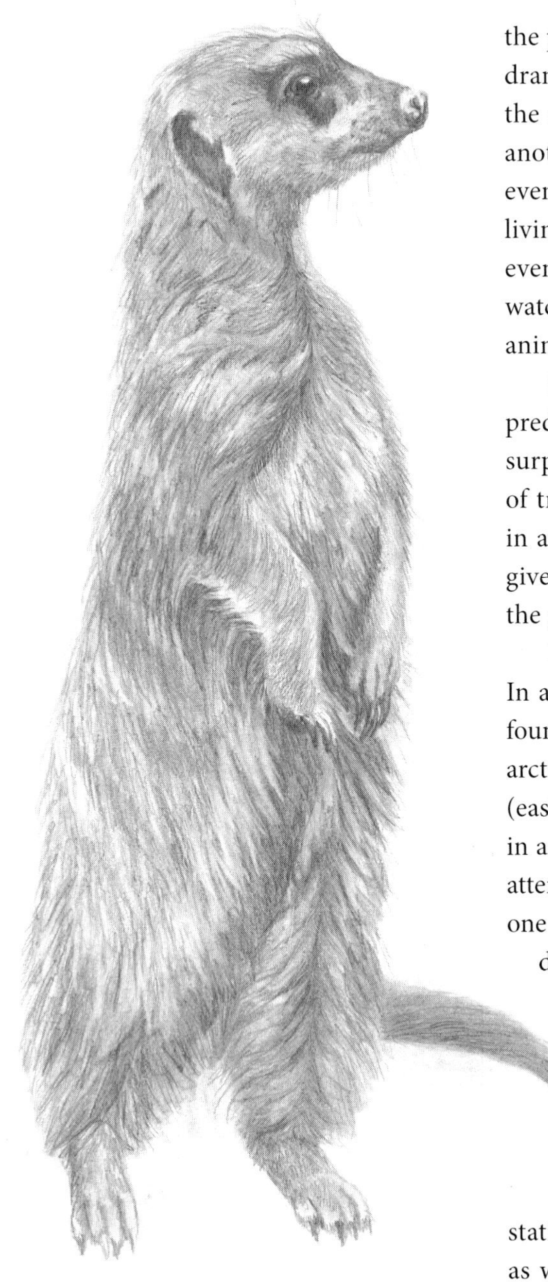

the probability of advance warning increases dramatically. While you yourself may miss the approach of a predator, the chances are another member of the group may notice even if you do not. Indeed, in some group-living species, such as meerkats, the group even posts 'sentinels' who concentrate on watching out for possible danger, while other animals may concentrate on foraging.

Since successful attack by many predators depends heavily on the element of surprise, this early warning system may be of tremendous value to group-living species in avoiding attack in the first place, but also gives sufficient time to allow the members of the group to run for cover or concealment.

Nor is this just comfortable theorising. In a study published in the early 1970s, Carl found that he could creep up on individual arctic ground squirrels to within three metres (easily close enough for a real predator to rush in and make a successful kill), while when he attempted to approach groups of individuals, one or another member of the group would detect him and sound the alarm while he was still as much as 300 metres away. Anyone who has watched wildlife programmes on the television must be well aware of other similar examples.

There is a further, purely statistical, element of this safety in numbers as well: the more of you there are (whether the predator was undetected, or whether the group is now in flight), the smaller the probability that it is you that is going to get caught. If you are on your own, the predator is definitely gunning for you. If you are one of a group, there is an increasing likelihood that the predator will target another individual in that group. While this may seem a rather nebulous advantage (particularly given our earlier suggestion that animals in groups are more likely to be detected by a predator in the first place), there may be an additional advantage in numbers if the early warning system was effective: in that, where all group members are running to escape, there may be some element of confusion for the predator in selecting an individual for attack. In

In groups of Arctic ground squirrels, someone will sound the alarm when a predator is as far away as 300 metres

most cases predators are at their most efficient if they can home in on a given individual and maintain pursuit of that individual. This is perhaps quite obvious in relation to those predators which rely on wearing their prey down by prolonged pursuit; if such predators do not maintain that pursuit of a tiring prey, but switch attention halfway through the chase to another, fresher, individual, their chances of making a kill are much reduced. But the principle is much the same for those predators who work by smash-and-grab: if they, too, keep switching their attention from one individual to another, it is much harder to make a kill at all.

Once again, this is now common experience from wildlife television documentaries. Groups of fish surprised by predatory species form themselves into tight and rapidly swirling balls within which each individual on the outside constantly seeks the safety of the centre of the mass. The overall effect is of a glittering, swirling mass, and it is extremely difficult for the human observer to keep an eye on any one individual. Hunting lions, or cheetah, rely on singling out an individual antelope or gazelle and separating it from the rest of the herd; much of the hunting strategy, indeed, is designed to achieve this separation of a single individual from the rest, for their success in maintaining pressure on any individual and making a kill within a churning milling herd is similarly enormously reduced.

While, in this last example, the benefits arise in some measure from confusion within the melee, there is also a potential within the group for effective defence against the predator through actual counter-attack. Some prey species may make pretty formidable opponents; few lions are successful in killing a male African buffalo in its prime, even if they come across it on its own but, faced with a group *en masse*, they are even less effective, and even those species or individuals which may lack the weapons or strength to ward off attack individually, may provide more effective opposition as a group.

Adult male African buffalo on their own may be more than a match for most lions – which is why the lions focus their attentions on younger females and, particularly, calves which have far fewer defences against attack. But when they are threatened by lions, and if they have detected them well enough in advance, groups of buffalo form themselves into a tight circle, slowly turning always to face the threat. Females and calves are packed within the centre, and mature bulls line the circumference, facing outward with lowered horns and presenting a formidable barrier to the lions. Groups of musk-oxen in the Arctic form the same defensive ring when threatened by wolves.

And group defence may even extend to counter attack. Baboons are favoured prey for leopards, who seem particularly partial to them. Individual baboons are relatively easy prey, and even a large male is not much of a match for a leopard. But if a larger group of baboons is attacked by a leopard troop, while the females and young scatter and run for safety, a detachment of mature males may break away from the fleeing troop and not only face up to the leopard, but run forward in attack. Male baboons have extremely large, slashing canine teeth, capable of inflicting serious injury; few leopards will face a posse of them. The massed counterattack effectively turns the tables and I myself have seen many a leopard turn and run.

Increased foraging efficiency

Living together with others may also offer advantages in greater efficiency of foraging. Although the down-side is that any food found must be shared, there may be situations where an individual's net food intake is nonetheless increased by foraging as a member of a group. Some of this benefit may derive from sharing the burden of looking out for potential predators, as we have just discussed. If there are more eyes and ears around you on the alert, you as an individual may be able to afford to spend slightly less time overall checking for danger and may be able to spend more time searching for food; if like meerkats you actually take turns on sentry duty, so that when it is your turn you do nothing else but scan for danger, when you are off-duty you may not need to bother at all, but may spend all your time feeding. Either scenario may increase the overall time available to search for food and consume it.

As a foraging group, you may also have greater competitive ability, in being able to commandeer food patches once discovered – even perhaps displacing other groups or individuals from favoured feeding patches – and in turn in defending ownership of that patch against intruders. In addition, it may be possible to exploit types of food, or particular food distributions that would not be available to you as an individual. Olive baboons, for example, frequently live in arid semi-desert areas of Africa where patches of fruit or insect prey are very patchily distributed. Food patches are highly aggregated: there are vast tracts of ground with little or no food, but when a patch is encountered, it may be extremely productive. An individual foraging in such a landscape has a very limited chance of encountering any food patches at all. If it does so, the ephemeral nature of the food (insects which can escape, fruit which can rot) means that it is unlikely to be able to exploit all of the food it has found anyway. If a group of baboons spreads out in a searching line across the ground, there is a far higher chance of encounter with a food patch; sharing that food patch with others subsequently isn't so much of a problem because, as we have just remarked, there is likely to be more than one individual can use in any case – and because even after that patch is exhausted, the frequency of encounter with food patches by the searching group is relatively high, so that there is a high likelihood of some member of the group finding another food patch in the very near future. Once again, individual intake rates may be higher in the group than in foraging alone.

Adaptiveness in this regard, and how the balance of advantage/disadvantage of group-foraging depends on the precise distribution of food, is elegantly illustrated by studies by Henry Horn of the American Brewer's blackbird. Horn found that in environments where food items were reasonably evenly and regularly distributed in space, blackbirds foraged as

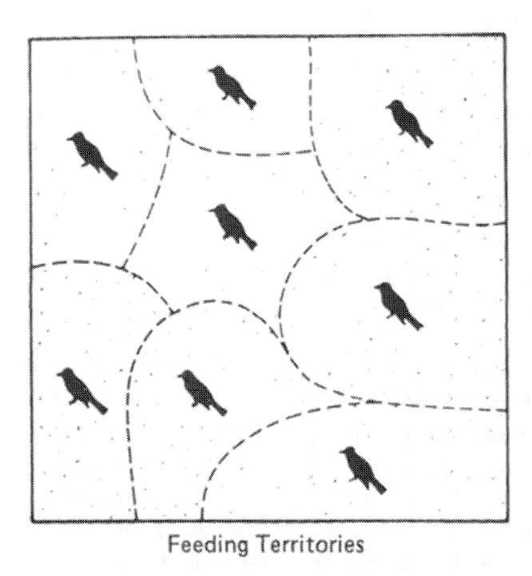

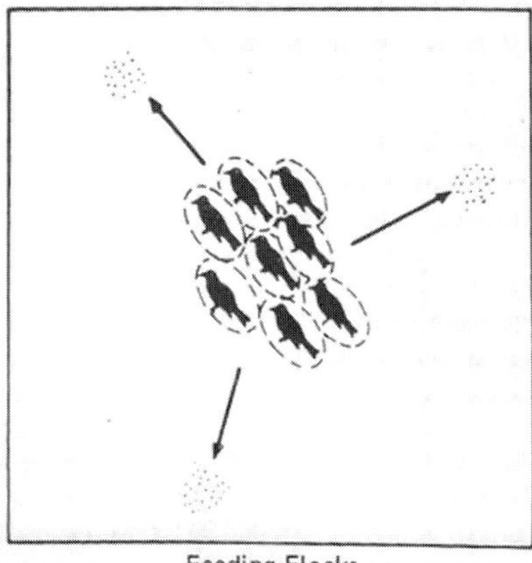

Henry Horn's analyses of group foraging in Brewer's blackbirds. If food is more or less evenly distributed in the environment and can be economically defended, it is most efficient to occupy exclusive territories. However, where food occurs in unpredictable patches, it becomes more efficient to forage as a group. (Horn, H.S. 1968 *Ecology* 49: 682–694)

solitary individuals and, indeed, even defended personal territories against other individuals to protect their food supplies. In other places – or in other seasons – where food items became aggregated, with available items concentrated but patchily distributed, as we have just described for baboons, the blackbirds abandon their individualist approach and form into feeding flocks which together range over larger foraging areas.

Group foraging may also facilitate the exploitation of food types or hunting methods which could not successfully be hunted or utilised by an individual on its own. River dolphins may use teamwork to drive schools of freshwater fish into the shallows, or even onto the river bank, where they are trapped and may be more easily picked off; feeding pelicans may also surround a school of fish in a tight, co-operative ring, to trap them and prevent escape.

And groups of lions may co-operate to flush antelope or gazelle towards an ambush prepared by other members of the group, again increasing their hunting efficiency. Data collected by George Schaller in the Serengeti showed clearly, for example, that if hunting either zebra or wildebeest, the proportion of chases which resulted in a successful kill was only 15% for a solitary lion, 31% for pairs of lions hunting together, and rose to 43% for groups of lions of between six and eight individuals.

These data are worth investigating in a little more detail, for they demonstrate very nicely that group living has disadvantages as well as advantages and that there is a balance to be struck between the two. Schaller's original data were explored in some detail by Thomas Caraco and Larry Wolf and we will rehearse some of that further analysis here. Caraco and Wolf noted that, amongst pride lions, hunting groups varied considerably from season to season and in relation to the type of prey hunted. Whether group size was determined by prey availability, or whether the choice of prey to hunt was determined by group size, it was clear that particular prey species tended to be hunted in groups of a particular size. Caraco and Wolf argued that group size would affect the size of prey which could be handled in the first place (with some prey types too big to be hunted by only one or two lions), the actual efficiency of capture and the efficiency with which the lions would be able to defend their kill against scavengers such as jackals or hyaenas. On the other hand, the larger the group size, the less meat is available from any kill per individual lion, since the kill must be shared - and for our calculations here, we should note that an adult lion requires, on average, an intake of at least 6 kilos of meat per day.

Let us take two examples and try to predict from this sort of consideration the optimum group size into which lions should divide themselves when hunting a particular prey species. One of the common prey species of lions in the Serengeti is the Thomson's gazelle. These little antelope weigh on average about 15 kg and some 75% of that is actually palatable to lions. Thomson's gazelles offer only a small carcase, which may be consumed very rapidly, so losses to scavengers are minimal and will not be greatly affected by group size. As we have already realised, not every chase leads to a successful kill (in fact, lions are not amongst the most successful predators); when hunting Thomson's gazelle, a single lion has a success rate of 15%; two lions have a success rate of 31% and a further increase in group size does not increase efficiency beyond this: larger groups of lions still only kill a gazelle in 31% of attempts.

From such statistics we may calculate the actual edible weight of meat per lion which will result from kills by hunting groups of different size as:

average carcase weight × proportion edible × capture efficiency × proportion saved from scavengers × number of lions in the group

In the case of Thomson's gazelle, we find that the maximum return per individual lion in terms of edible meat is achieved when hunting in groups of two and works out at just under 2 kg per lion. Since the minimum daily intake required per lion is 6 kg, we might predict that when hunting Thomson's gazelle, lions should hunt in pairs and should hunt on average three times a day. In fact, Schaller's original data show that three-quarters of all hunts of Thomson's gazelle involved single lions or pairs of lions, with an average group size very close to two. And when relying primarily on Thomson's gazelle, lions tend to hunt three times a day.

Let's try this again for lions hunting zebra or wildebeest. Here, we are dealing with a larger prey item, with an average weight of 180 kg; 62.5% is actually edible or palatable to lions. The carcase size is such that the risk of losses to other scavengers is real, with 10% of the carcase (or 10% of carcases) lost where the hunting group numbers fewer than four lions. We have already noted that for such prey items, capture efficiency is 15% for a single lion, 31% for pairs and 43% where six or more lions are involved in the hunt. With zebra or wildebeest, there is also an additional factor in our calculation: that given the carcase size, it will last a feeding group on average three days, whatever the size of the feeding group. Using the same formula as before, we may calculate the amount of edible meat per lion per day for lions in groups of different sizes. In this case we find that groups of two lions achieve the greatest prey intake rate (at 9 kg per day), but that groups of between one and four lions still achieve the physiological minimum requirement of 6 kg per lion per day.

In practice, Schaller's original data show that lions tend to hunt zebra and wildebeest in groups larger than we might predict on grounds of maximum foraging efficiency alone, with an average group size in fact of between four and seven lions. While perhaps maximum group size is constrained by energetic considerations, actual group sizes appear larger than would be optimal simply in relation to feeding efficiency, suggesting that other factors too may be important in this case in promoting sociality.

Social advantages of living in groups

Caraco and Wolf suggest that the fact that group sizes of lions hunting large prey tend to be larger than we would predict from energetic considerations alone may be due to other advantages of sociality, such as group defence of cubs, co-operation in the rearing of cubs and so on. (Lionesses in a group not uncommonly give birth at around the same time, they not uncommonly permit other females' cubs to suckle, and regurgitation of prey to growing young is also fairly indiscriminate within the group).

This introduces another suite of possible advantages to living in a group, which are commonly referred to as social advantages, including regular access to the opposite sex, co-operation in breeding and in the rearing of young, and the presence of peers during the period of social learning.

Why don't all animals live in groups - and what determines actual group size?

While there may be a number of potential advantages, in some circumstances, to living together in groups, such sociality clearly also has its costs - and not all species do choose to live in groups in this way. What factors, then, will determine whether it is better for a given species to be solitary or social? And even if a species is better served by being social, what determines actual group size?

In broad generality, it would appear that optimal group size is closely related to ecology, particularly to habitat and food distribution. The group sizes adopted by Serengeti lions depended in large part on which particular species of prey they were hunting, while the Brewer's blackbirds studied by Horn gravitated together into feeding flocks when food was aggregated and patchily distributed in their environment, but operated as territorial individuals when food was more regular and uniform in distribution. The anti-predator advantages of living in a large group may outweigh the disadvantages to the individual in open country with little cover, where an early warning system and group defence may be the only effective strategy. In dense forest, however, a large group becomes extremely obvious and is, in any case, hard to maintain: perhaps the best strategy in this case is to escape detection in the first place, to be secretive, elusive and solitary.

Overall it is clear that to satisfy nutritional, anti-predator and social requirements from the particular range of resources on offer, group size does become precisely adapted to habitat type. The first appreciation of this close fit of group size to ecology was offered in the 1960s by Crook and Gartlan who simply categorised different species of primates by the type of habitat in which they lived and found a clear relationship between habitat and group size. Thus, dividing the different species into categories according to whether they lived in dense forests, forest edge, open tree savannahs, open grasslands or arid semi-desert habitats, they found a striking gradation of social structure such that (whatever their evolutionary relationship) monkeys which lived in dense forest tended to be solitary, insectivorous animals living in small home ranges, species which lived on the forest fringe tended to be fruit and leaf-eating forms which occurred together in small family groups, while species characteristic of open grasslands or arid areas, like our baboons of page 89, were characteristically vegetarian or omnivorous animals which lived in large troops, covering large home ranges as they exploited very patchy food distributions.

While the conclusions of a relationship between habitat and group size are fairly robust, there are in practice a number of flaws in Crook and Gartlan's analysis: habitat definitions are too broad and, in truth, there is nearly as much variation in group size within each category as between them. But the principles are sound enough. In a more detailed exploration of the relationship between group size and ecology among East African antelope, Peter Jarman showed a clear effect of both habitat and foraging ecology, such that for any species assuming a particular feeding style within a particular habitat, there was a particular group size to be adopted; almost irrespective of the species concerned, a particular group size appeared characteristic of a given lifestyle.

Animals like dik-dik, characteristic of dense forest, tend to occur solitarily or in mated pairs

	HABITAT TYPE				
	FOREST	FOREST EDGE/WOODLAND	SAVANNAH	GRASSLAND	ARID/SEMI-ARID (DESERT)
SELECTIVE BROWSER	Duiker (13 spp) singly or pairs Reedbuck (3 spp) singly or pairs	Kirk's dik-dik: singly or pairs Grysbok: singly or pairs	Steinbok: 2 – 4		
SELECTIVE BROWSER/GRAZER	Bushbuck: singly or pairs	Oribi: 2 – 6 Gerenuk: 3 – 6	Lesser kudu: 3 – 10		Lesser kudu: 3 – 10
BROWSER/GRAZER		Impala: 10 – 60 Waterbuck: 10 – 30 Greater Kudu: 10 – 20	Puku: 10 – 60 Kob: 10 – 60	Thompson's gazelle c60	Grant's gazelle 2 – 6
RELATIVELY UNSELECTIVE GRAZER		Topi 6 – 60	Jackson's Hartebeest Coke's Hartebeest		
UNSELECTIVE GRAZER	Buffalo: 2 – 6	Eland 60 – 100	Eland: 60 – 100	Buffalo: 100^{s} – 1000^{s} Wildebeest: 100^{s} – 100^{s}	Beisa oryx: 100^{s}

Relationship between social organisation and ecology: group size amongst East African antelope is related to feeding style and habitat (Data from Jarman, P. 1974 *Behaviour* 48: 215–266)

	Feeding style	Habitat				
		Dense forest/ thicket	Open woodland	Forest edge/shrub	Open grassland	Marshland/ carr
Concentrate-selector	Fruit-feeder Selective browser	Mazama: solitary/ pairs Pudu: solitary Muntjac: solitary/ pairs Huemul: solitary/ pairs Roe: solitary/ small groups		Huemul: <5	Roe: up to 40	Chinese water deer: 1–2 Moose: 1–2
	Browser/ grazer	Sambar: solitary/ small groups Hog deer: solitary/ small groups	Sambar: 3–5 White-tailed deer: 1–3		Hog deer: 10–20 White-tailed deer: 6–10	
Intermediate feeder	Grazer/ browser	Red deer: 1–3 Sika: 1–3	Fallow: 1–10	Chital: 5–10	Chital: 15–20 occasionally 100s Red deer: 1–50 Fallow: 30–100	
Bulk feeder	Grazer				Pampas deer: 5–15 Barasingha: 12–100s Rusa: 100s	Eld's deer: 'large herds'

Group size of different species of Eurasian deer, related to their ecology. (From Putman, 1988, *The Natural History of Deer*)

Amongst Jarman's antelope, browsers in dense forest were found universally to be solitary or to occur as mated pairs. As browsers they are highly selective in what they feed on in terms of types of food or parts of plants, but relatively unselective with respect to species. Such needs can be met in a small home range, but to restrict competition from other individuals, that range should be defended as an exclusive territory and occupied by the territory owner alone, or a breeding pair. Within dense forests, it is impracticable to try to maintain the cohesiveness of a large group, and thus the anti-predator advantages of such a group cannot be realised; in such dense cover, perhaps the best anti-predator strategy may in any case be to rely on avoiding detection in the first place rather than adopt a strategy of active defence, as we have already suggested. Once again, circumstances in this regard also favour a solitary habit.

By contrast, out on the open grasslands, relatively unselective bulk feeders are offered an abundance of food. Group size will not greatly influence feeding efficiency, either positively or negatively; while there is nothing to stop the formation of large groups in such an environment, nor is there anything about the type or distribution of food actually to promote the formation of groups. In these open habitats, however, grazing antelope are exposed and far from cover. The best anti-predator strategy, therefore, is to opt for all the advantages of early detection and group defence offered by being a member of a larger social unit – and there is no real cost in terms of reduced foraging efficiency. Anti-predator considerations will in these situations encourage the formation of larger groups in open habitats; there is nothing about food type or distribution which will similarly promote the formation of groups but, by the same token, there is nothing that would prevent such grouping. Intermediate examples may be explained by the same logic, and as a general rule we may observe that group sizes of antelope increase as the environment becomes progressively more open.

Differences between the social systems of different deer species may be accounted for on exactly the same basis; thus species such as brocket deer, or the pudu, of the South American rainforest are small browsing or fruit feeding species of dense forest and tend to be solitary; at the other extreme, it is amongst species of open grasslands such as the pampas deer, the barasingha or the rusa deer of Asia that we find the largest social aggregations. Species of open woodland and the woodland edge, such as European fallow deer or Indian chital, when they group at all, tend to be found in groups of intermediate size.

But herein lies the next subtlety: for this relationship between group size and environmental circumstance does not simply result in gross differences in group size between different taxonomic species, which are then fixed for that species. The relationship between group size and environment is more dynamic than that and we may see that among those species such as the roe deer, hog deer, fallow deer or chital, which are perhaps more flexible in their habitat requirements and can be found in a wide range of different ecological communities, the same variation in group size may be observed within the species, as each local population adopts the social structure appropriate to the habitat in which it occurs.

This within-species variation in social grouping is important evidence that these differences in social organisation are functional and adaptive, rather than simply chance taxonomic coincidence of species-specific social structures. Jarman noted this himself in realising that African buffalo of the open plains were associated in far larger groups than were buffalo of the same species occurring in forest habitats. Similar variation is shown amongst European roe deer in our second figure. Roe deer are relatively primitive deer which feed selectively on small morsels of highly nutritious forage: browse species or highly nutritious forbs and grasses. They are generally characteristic of dense woodland or woodland edge, but their small size and relatively modest food requirements, in terms of actual bulk, allow them to exploit quite a wide range of habitats and they may even be found in hedgerows and have even managed to adapt to the open agricultural landscapes of central and eastern Europe. Where they occur in the 'traditional' woodland environment, roe are found to be solitary or to occur at most in small family groups; in the agricultural prairies of central Europe, in perfect vindication of our theories, they may gather over winter in permanent social groups of up to 100 individuals. Similar responses are found among other species, such as white-tailed deer or fallow, where group sizes are typically much larger in populations characteristic of more open habitat than in areas where their ranges are predominantly within woodland.

This flexibility of adaptation – and the precision with which group size is matched to habitat – is further illustrated at a level of resolution one stage higher still. Among those species whose habitat is not strictly homogeneous, but offers a mosaic of habitat types patchworked together, group size does not even remain constant within any one population, but may change from day to day, even from minute to minute, as the animals move from one habitat to another, reflecting at all times the habitat occupied at that particular juncture. If permanently resident in one habitat type, groups will be of a constant size appropriate to that habitat, producing the responses we have just described.

In mixed environments, however, where animals may exploit a range of different vegetational types, group size will actually change as they move from one habitat to another. Fallow deer which live permanently in more open environments tend to live in larger social groups than do those whose range is predominantly within woodland; but where animals move between woodlands and open fields, groups coalesce and split again as they move from one habitat type to the other and group size constantly adapts to the present habitat.

Social organisation and ecology

So far we have been concentrating on the relationship between group size to ecology, and whether or not to form groups at all. But environmental circumstance has a direct affect on other aspects of social behaviour as well; indeed, we have already hinted at some of these in the past few pages. In relation to Jarman's analysis of social organisation of East African antelope, we have noted that small forest browsers, as selective feeders, will tend to be solitary and can maintain themselves within relatively small home ranges, but that in order to do so effectively they should defend these home ranges against other conspecifics. If they are to maintain themselves successfully within a small range, they must ensure uncontested access to the all resources contained within that range and thus must actively defend it as an exclusive territory. By contrast, the unselective grazers of the open grasslands, mere unchoosy mowing machines, can afford to form large groups in the open, but in consequence may rapidly exhaust the forage resources of a given area. This means that such animals can have limited fidelity to a permanent home range and this, coupled with the seasonal nature of the East African climate, means that most such species are in fact forced to become seasonal migrants.

Such considerations may be extended more generally. Whatever the group size, considerations of the predictability and quality of food supplies may determine an animal's ranging behaviour and the degree to which it may tolerate others on its patch. For once we will illustrate this by human example, from the elegant analyses by Rada Dyson-Hudson and Eric Smith on the social organisation of aboriginal Amerindians. It would appear that social structures of humans can be relatively simply explained in terms of two characteristics of their environment: the density and predictability of food resources. Where resource density (patch quality) is high and predictability is high, tribal peoples tend to be sedentary and

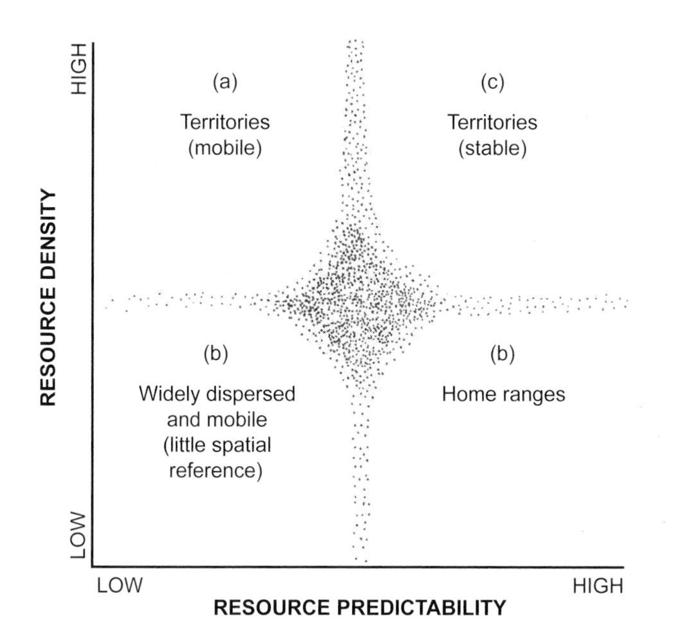

Dyson-Hudson and Smith's schematic representation of how human hunter-gatherer societies respond to abundance and predictability of resources, redrawn from Dyson-Hudson, R and Smith, E. 1978. Human territoriality: an ecological re-assessment. *American Anthropologist* 80: 21–41 and reproduced by permission of the authors and the American Anthropological Association.

territorial. Where resource predictability is still high but quality or abundance is poor, then such aboriginal peoples tend to occupy large home ranges, which may overlap extensively with those of other groups. Where resource quality is high but predictability is low, defence of resources is still appropriate within established territorial boundaries but, due to the low predictability of rich patches, these are defended wherever encountered and thus territories are not fixed in space. Where resources are of both low density and low predictability, these may be exploited only by highly mobile peoples, with overlapping ranges, and there is no value in defending resources, which are exhausted as soon as encountered.

Ecological factors may thus affect group size, may affect the 'decision' on whether or not to be territorial, and whether or not to be sedentary or migratory in habit. Ecological circumstance may also affect the permanence of social groups themselves and their internal structure.

Let us return to our exploration of the variation in group size in deer in relation to their ecology. We have noted already that we can explain the observed differences between species in terms of habitat occupied and foraging strategy; we have noted, too, that we may detect similar variation within species, where different populations may occur in habitats of different structure, emphasising that the phenomenon is driven by immediate ecological circumstance and is not simply an evolutionarily fixed, species-specific character. But group size does not simply differ between different local populations of a given species; even within a population, group sizes may change as individuals move from one habitat to another within a mixed environment. This has considerable implications for the type of social structuring that may develop within the group.

Red deer in woodlands are commonly encountered individually or in pairs; in mixed environments, where they may move between a variety of different habitats, groups form and break. When in dense woodland, the deer still operate essentially as individuals, but when they feed out into clearings, glades or woodland rides, they group together into larger assemblages - forming larger groups still when they forage beyond the woodland edge into open moorland or agricultural fields. Individuals join together as they move into more open areas; groups fragment again as the animals return to denser cover. It is the

habitat of the time that determines the tendency to group and the size of group formed, and both group size and cohesiveness are extremely flexible.

Such flexibility and change is not, however, universal. In the north of Scotland, the same species has become adapted to living year-round on open moorland. Individuals come together in sizeable groups, but, since the animals do not move between this open habitat and others, this larger group size remains continuously appropriate, and there is not the fission/fusion so characteristic of populations in more diverse habitat mosaics. Indeed, in such situations, red deer have been found to form relatively permanent and persistent social units, based on persistent hierarchical, matrilineal groupings.

In this way, in different ecological situations, we may observe the development of different social structures even within the group. When the deer live in open habitats where, year round, larger group sizes are appropriate, groups formed may become relatively permanent family groupings, whose membership is fixed – and because of the permanence of individual membership, such groups may develop quite complex social structures and clear dominance hierarchies amongst their members. But where red deer live in more mixed environmental mosaics, groups constantly form and disintegrate as the animals move from one habitat to another. Even when groups reform, they will not necessarily be composed of the same individuals on each occasion. Here, where the composition and individual membership of groups is constantly changing, there is little opportunity to form persistent associations between individuals; groups have no real structure, and within them the basic social unit remains each separate individual.

Thus, in the same way that a shift from closed to open environments may be responsible for a change from solitary to more aggregational habit, differences between the more social species (and within species) in the permanence of association and thus degree of sociality may also reflect environmental circumstance, although in this case relating more to the homogeneity or heterogeneity of the vegetational mix in which the particular species or population is found. Temporary aggregations within which each animal remains essentially individualistic will be characteristic of environments offering a complex mosaic of open and closed community types; more truly 'social' groups of permanent membership, within which may be developed more complex social structures, will be found in environments both open and relatively homogeneous.

Eusociality

In all the examples we have considered so far, even though they may associate together into large, even permanent groups, each individual organism retains its individuality within the group. Perhaps the ultimate extreme in living in groups, though, is the formation of colonies by some animals where individuality is to a greater or lesser extent suppressed and individual colony members become in some sense integral members of some 'superorganism'.

Colonial development of this sort is clearly at its most advanced among invertebrates such as social insects (like bees and wasps or termites) – although there are some striking examples among vertebrates, such as for example, the various species of African mole-rats. In such societies, (which are sometimes referred to as eusocial), the existence of the individual is suppressed within the life of the colony. Usually, there is some degree of reproductive inhibition such that there is commonly only one reproductive female: the colony founder, or 'queen'. The majority of other individuals are non-reproductive, and may become highly specialised to particular functions within the colony: as foragers, nursemaids or 'soldiers'. Indeed, it is frequently the case that any one individual is unable to exist on its own: it has become so specialised to serve one particular function within the colony, and to have its other needs served by other specialists within the colony, that it is physiologically unable to survive alone.

Thus it may truly be argued that the 'individual' within the colony becomes almost a component organ of some superorganism. In many colonial coelenterates this is literally true, in fact, in the sense that different individuals are physically fused to each other; the float, stinging tentacles and reproductive 'organs' of the Portuguese Man-o'-War for example, while they appear part of a single whole, are technically different specialised individuals of a colonial association so tight that it even resembles a single structure.

Even among social insects, the interdependence may become so tight that it is better to regard the individuals not as independent entities. Such an argument is further strengthened by the fact that in almost all such cases (including mole-rats!) the individual members of the colony are very closely related to each other or even genetically identical. In that sense a clonal colony of genetically-identical individuals may indeed be regarded in its own right as a single entity, rather than as a grouping of distinct individuals.

In any event, it is clear that in this extreme sense – where colonial organisation involves the integration of a number of genetically-identical members, reproductive suppression of all but a few, and extreme physiological specialisation of most non-reproductive individuals – we are dealing with something rather distinct from the sorts of social groups we have been discussing so far and there is considerable controversy as to whether they may be treated within the same conceptual framework as other forms of sociality. Some argue that such colonies are merely an extreme on the continuum from solitary through social organisms – that the suppression of individuality within the group is merely an extension of the more general trends to be found in increasing sociality in general. Others argue, as I have here, that while they are indeed simply an extreme development within a broader evolution of increasing sociality, their very structure and the general suppression of individuality within the whole render such eusocial societies something of a special case: that such societies have become so specialised in structure that they can really no longer be considered as social groupings of independent individuals, and must be treated as a separate special category of their own.

Reproductive behaviour and reproductive strategies

Basic life-history strategies and how to make the most of your reproductive investment

An essential basis of life is that organisms should reproduce themselves. And indeed the whole basis of evolutionary adjustment to an organism's basic structure, physiology, behaviour, is to improve its reproductive success – since evolutionary change (or stability) depends in effect on differences between individuals in the number of viable offspring they may leave to succeeding generations. Natural selection thus acts to produce animals which are adapted in every regard to maximise the spread of their own genes, but just as we discovered in relation to social organisation, there is more than one possible solution to the problem, and just as environmental circumstances affect whether it is better to be social or to be solitary, so the 'choice' of reproductive strategy adopted by any animal species is closely related to ecological circumstance.

In practice, at the species level, there are two alternative strategies which may be adopted. A species may simply 'opt' for producing the maximum number of progeny that it can. Given, however, that animals have limited energy budgets and thus can literally afford only a finite given investment into reproduction, if all the available commitment is invested in production of the maximum number of offspring possible, nothing is left in the pot to offer any care to those offspring – and mortality rates may be high. A viable alternative then is to opt for relatively low reproductive recruitment, in the production of relatively few offspring, but ensure high survival of those offspring that are produced by investing heavily in parental care.

These two alternatives are mutually exclusive. Since there is a necessary trade-off between the number of offspring which may be produced and the level of parental care which may be offered to those offspring, then the greater the investment in parental care, the lower the number of offspring that can be produced. Given that one can only produce a small number of offspring in the first place under this strategy, it is then crucial to ensure maximum probability of survival of those few offspring by maximising the investment in parental care – further compromising the number of offspring you may produce in the first place. Conversely, the higher the number of offspring produced, the lower the possible investment in each in post-natal care; survival of the young of more prolific species may thus be greatly reduced. Since low survival can only be compensated for by large initial numbers, evolutionary processes will act to accentuate the strategy, reducing parental care still further, maximising recruitment. Thus the two basic strategies must polarise at extremes: many offspring and little care, or few offspring, supported by considerable parental care.

Either method is valid; each is a perfectly viable option and different species 'adopt' one or the other depending on their circumstances. In practice, the two strategies have a different balance of advantage or disadvantage under different conditions: high fecundity and rapid development, with little or no parental investment, is the better strategy to adopt in rapidly-changing or relatively unstable environments –'unstable' in either space or time. Carrion blowflies, for example, breed of necessity in resources which are spatially scattered and are themselves ephemeral. The best way of ensuring continuity under such conditions is to maximise production of young while you can, producing the maximum number of dispersing adults so that even though many may die, some at least may find the next carcase and continue the cycle. By contrast, organisms characteristic of relatively stable environments tend to have their population sizes closely adjusted to the available resources (the 'carrying capacity' of that environment). Such organisms can recruit only as many surviving young as are sufficient to replace annual losses within that stable population; to produce any more would cause the population to exceed the capacity of the environment to support it, would risk damaging that environment and consequent high mortality. To ensure that those few offspring survive, however, they must invest heavily in their care. Thus we find different organisms specialising in one or other basic strategy, dependent in large part on the stability of their surroundings.

Breeding systems: monogamy, polygamy or promiscuity?

Within such basic choice of bionomic strategy, however, there is a huge potential variation in the breeding system which may be adopted. For species which are essentially adapted to maximise simple production of offspring, mating systems are generally promiscuous. But, within that class of organism that opts for some degree of parental care, we find a whole range of possible breeding systems from monogamy, through polygamy, to total promiscuity. We may find polygamous animals simultaneously maintaining numerous partners, or being completely 'monogamous' with a series of different partners in quick succession (serial polygamy); sometimes one male is associated with many females (polygyny); in other instances one female is associated with many males – again, either simultaneously, or sequentially (polyandry). Since, once again, it is apparent that all these different systems are viable, what factors determine which of these breeding systems is adopted in any given instance?

In essence, in the same way that for a species a given strategy has evolved to maximise long-term reproductive output in given circumstances – through maximising production of offspring, or by ensuring through parental care the maximum survival of relatively fewer offspring – so the same selection pressures continue to act on all individuals within the species. And the basis for the adoption of one or another mating system, from monogamy to promiscuity is that the selection pressures do not act in the same way on both partners in a reproductive attempt. That strategy which might optimise reproductive success for a male in any given species, for example, might not be the strategy that would maximise reproductive success for the female. The eventual system adopted is, if you like, a compromise between the pressures acting on the two partners in the breeding attempt.

Natural selection acts at the level of the individual, not the species. At any stage, therefore, each individual 'wishes' to adopt that strategy which maximises its own production of young (and minimises cost). The options are very much as before: if you can decrease the investment needed in parental care (defending a territory, guarding a female or young, bringing food to a partner or growing youngsters) then you can increase investment in producing more young in the first place. If you decrease the production of young, you can increase the investment in parental care. But what is important here is that the investment generally differs between the two partners in any breeding attempt.

At any stage, as we have said, each individual is under heavy selection pressure to adopt that strategy which maximises its own production of young. In consequence, there is a logic which says that, at any point in time, the individual whose total investment is exceeded by his partner's may be tempted to desert. (More accurately, the partner who has invested more, who would thus have to invest more in recommencing a new breeding attempt but has by virtue of its earlier heavier investment, less left for such future investment, is the partner less likely to desert any breeding attempt). This temptation arises because the deserter loses less than his partner if no offspring are raised and the partner will be therefore more strongly selected to stay with the young. Any subsequent success of the remaining partner is of course of benefit to the deserter, but meanwhile he has another chance of success in another breeding attempt as well. For example, desertion by the lesser investor right after copulation will cost very little, even if as a result no offspring are raised from that mating; the chance that his partner may on her own successfully raise some of those offspring without his further intervention may be great enough to make the desertion worthwhile.

More generally, if any animal has to invest comparatively little in any individual breeding attempt this has two significant consequences. First, it has the potential to invest in a good many such attempts – and is in effect limited by the number of mates it can find, rather than by resources available for reproduction. Secondly, as we have just said, simply because it invests little in any individual breeding attempt, selection will have acted on the opposite sex (if it is to achieve any reproductive success at all) to be able to manage to rear any offspring on his/her own, so that in effect, the survival of offspring from a one-parent family is at least greater than zero, and possibly not much less than were the family cared for by two parents.

Under such circumstances, as long as there is a realistic probability of encountering another mate, the lesser investor is theoretically tempted to become promiscuous: copulate and desert in search of future copulations. For even if the survival of one-parent-reared offspring may be less than the chances of survival within a two-parent family, and even if he only gains one other mating (and stays with that second partner), as long as some offspring may survive from the first (deserted) mating, this desertion for a second mate will still increase overall the number of offspring he leaves to posterity since it will be the sum of the normal full complement from the two-parent reared second family, plus the bonus he may gain from any survival of offspring from the deserted family, reared by his first partner alone. As long as that bonus is non-zero, it will pay the individual making the smaller initial investment to desert and try his luck elsewhere.

By contrast, any animal which has invested heavily in a given reproductive attempt ought to stay with that attempt. It is not a simple question of the fact that having invested heavily one should not 'waste' that initial investment, but once committed should continue to invest regardless (an attitude rightly referred to among biologists as the Concorde Fallacy in honour of the same approach adopted in the development of that supersonic aircraft). Rather it is because the individual would have to make that same investment all over again in any future breeding attempt to get to the same point and simply may not be able to afford to. This individual is thus resource-limited, rather than mate-limited. Clearly, if some offspring are likely to survive even if she continues the breeding attempt on her own, she should continue with the attempt and remain with the young even after desertion. The investment-limited partner must stay. But she is losing out. If the actual survival chances of a two-parent family are greater than those she rears alone, it is in her interest to put pressure on her partner to stay and increase her reproductive success. But she can only do that by making 'staying' in his best interests too, by offsetting any potential advantages to him of desertion.

She may do this in a number of ways. Firstly she can increase the investment he must put in to the early stages of even the first breeding attempt (investment both of resources and of time). She may do this by refusing to mate unless a male has established and is actively defending a territory, or she may play coy, 'insisting on' a long courtship. Such demands increase his investment so that he cannot afford so easily to abandon any breeding attempt and repeat that investment all over again – so that in effect, like herself, he becomes resource-limited rather than mate-limited. Such devices may also waste his time, so that if breeding is seasonal and highly synchronised within a population, the time demands of these requirements for a heavier initial investment also reduce his chances of finding other receptive females if he were to desert. And as long as all females in the population play the same game, then they can 'force' these demands upon the male, such that he will have to make these heavier investments if he is to have any chance of breeding at all. In such a case, this greater required investment ensures that now his best reproductive interests are also served by ensuring the maximum survival of the one breeding attempt he may achieve, and thus he too will vote for staying and rearing any offspring within a two-parent family. Ultimately, his investment may even come to exceed hers, in which case she is now theoretically tempted to desert and start again.

Finally, the greater investor may turn the tables by making her partner the 'terminal investor', making his investment follow hers, so that she has a chance to escape and literally leave him holding the baby(s); perhaps the female stickleback chased from the male's territory after fertilisation of her eggs, doesn't get such a raw deal after all? The actual system adopted in any species – monogamy, polygamy, promiscuity – reflects differences in the relative breakpoint between the optimal strategy for the two sexes and the degree to which the lesser investor can be manipulated to prevent desertion.

In general, my relapse into use of "his" and "hers" here is entirely appropriate, for in most animal species – and certainly among that class of animals which offer some degree of parental care to their offspring – it is indeed the male who makes the smaller initial investment. His contribution to any mating attempt is simply a small package of sperm, while the female contributes a much more significant egg; so that from the word go, there is a disparity in investment.

[As an aside, we should note here that this argument is often, but erroneously, justified on the basis of a difference in relative energetic cost of a tiny sperm and the much larger egg cell. However intuitively easy to grasp, this logic is in fact flawed, since sperm do not tend to come singly, and one ejaculate, together with all the nutrients contained within the seminal fluid, may be energetically at least equivalent to an egg. In practice it is not so much the energy value of the gamete contributed which is significant, but the supply. Males, of almost whatever species, vertebrate or invertebrate, continue to make sperm throughout their life; the germinal tissue of the testis is composed of sperm mother cells which actively and regularly produce fresh sperm cells, so that the supply is theoretically endless and each ejaculation does not threaten any future breeding attempts. By contrast, in many species (certainly amongst mammals and birds and also in a number of invertebrates) a female's supply of eggs is limited. All the egg cells available to her are present at birth and her germinal tissue does not continue to make new cells beyond that point. Instead, her production of eggs for fertilisation is through gradual maturation of the undeveloped eggs already present in the ovary from the point of birth; for her, each ovulation represents a finite proportion of her lifetime reproductive potential and each egg matured and shed effectively reduces her remaining reproductive capacity. While, in truth, each female is provided from birth with many more eggs than she probably ever will use in life, the supply is nonetheless finite and thus her contribution of an egg to any reproductive attempt is a far greater investment than a package of sperm.]

This initial disparity may be increased if the female's subsequent investment continues to be larger than the male's (as, for example, in internal gestation of the young through pregnancy) or may be overridden if the male is himself required to invest more than simply sperm in the initial stages of the breeding attempt. If, as we have suggested, the female's breeding tactics are such that she requires from the male a defended territory before copulation, requires him to contribute to the building of a nest or to contribute in some other way, the difference of investment may be reduced or cancelled out. Indeed, in extreme cases, a female may demand such an investment from the male during the early stages of a breeding attempt (an investment, which as long as every female of the same species plays by the same rules, he must make if he is to have any breeding success at all) that his investment comes to be greater than hers – and she is then able to desert.

Where males and females show similar investment in their offspring (or can be forced to make equal investment), then there will be a tendency for breeding systems to be monogamous, although simultaneous polygyny (a male having multiple wives) may occur if the advantages to the female of mating with an already-mated male and sharing his attentions with other females are nonetheless greater than those of mating with an unmated male. (This might occur, for example, where the already-mated male owns a territory which is far more food-rich than that of his unmated neighbour; even though the female must share the male's attentions – and the food resources – with other females if she mates with him, her share of the shared resources may still be greater than she would get from access to even 100% of the resources of the poorer territory).

In species where the male investment remains much less than that of the female, despite any efforts she may make to get him to increase that investment, we might expect to find polygamous mating systems: the female is going to bring up the offspring with reasonable success with very little additional input from the male, so the best way of maximising his overall reproductive output is to rush off and find some additional mates. Where, in the extreme, the male investment is purely in copulation, the likely strategy will be open promiscuity.

If a female is in some way able to increase the male's initial investment to match her own and then require still further input so that her investment is exceeded by that of the male, we might expect to find by converse, polyandrous systems, where one female, simultaneously or sequentially mates with many males. Such systems may also develop, as we have mentioned earlier, if the female can engineer that the male makes his investment (however small) follow hers, so that he is left literally holding the baby.

And in practice, the different breeding systems we may observe within the animal kingdom neatly reflect the degree to which the female is able to balance out the initial disparity in investment. In the vast majority of fish species, eggs and sperm are merely shed into the water (or if there is internal fertilisation, fertilised eggs are simply shed) and left to their own devices. There are a few species which show some degree of parental care (such as mouth-breeding cichlids, seahorses and our old friend the stickleback) and it is notable that in many of these it is in fact the male who cares for the young. Amongst fish, there is far less disparity between the sexes than for many other vertebrates, even in the initial investment in gametes, so balancing the investment is relatively easy; furthermore, with the breeding system adopted by many species, the male is commonly the terminal investor, so that male care is not uncommon.

Amongst birds, the initial disparity of sperm and egg is further increased. In this case there is a real energetic consideration, too, since a female's eggs are so very large and nutrient rich and she may well be limited in how many she can lay simply by available resources. But she can relatively easily force the male to increase his initial investment too, as we have seen, by establishing a territory, helping select a nest site, contributing to the building of the nest (in some species, the male actually builds the nest alone). Beyond that stage, there's actually not that much that the male cannot be required to do equally as well as the female. A male can incubate eggs just as well as a female can, so that they may take equal turns; if the female does 'elect' to do all the incubation, the male can be required to forage for her and provision her at the nest. Both male and female can work equally to protect and feed the hatching chicks and care for them until they reach maturity. So that as long as the female can overcome the initial large disparity in gametes, investment of both partners can be kept at level-pegging thereafter.

In consequence, most birds are monogamous. There are a few species which are polygynous – either because the females of the species have not adopted tactics to force the male to increase sufficiently his initial investment to match their own – or because there are sufficient differences in male quality (or quality of the resources they command) to make it worth a female's while sharing a quality male with others rather than have the sole attentions of a poorer quality male. Such polygyny has been noted for example in some (but not all) populations of hen harriers, and it has been seen to occur most often where there is a high density of nesting. It is remarkably developed in Orkney, where it was first noted in 1931, and where in recent years the sex ratio has been consistently two females to one male, with some males on the best territories supporting up to six females simultaneously.

Because, as we have noted, there is in fact relatively little that a male cannot do just as well as a female after the eggs are laid and so she can force him to increase his investment to match or even exceed hers, it is not surprising to discover that amongst birds – just as with fish – there are also occasional examples of a reverse polygamy, with one female mating, usually sequentially, with a number of males.

Female hen harrier at nest

A classic example of this is provided by Temminck's stint, where a female will pair with a male, establish a nest and lay her eggs, only to abandon the nest at that point and repeat the overtures with a new male on a new territory; the deserted male incubates and rears the chicks on his own. One individual female may lay eggs for up to three or four males, abandoning each in turn until she finally settles and stays with the final male of the series and shares the burden of bringing up the final batch of offspring.

While other groups may balance the odds or even reverse them, amongst mammals the disparity in investment is too great. There is a large initial disparity in the 'value' of gametes, which is compounded by the fact that mammals are mammals: that is, the foetus is kept internally by the female and nourished directly via a placenta throughout its development – and even after it is born, only female mammals produce milk. As a result of the female-only aspect of placentation and lactation, the difference between the sexes in investment in reproduction is enormous. Not only that: the same phenomena of female pregnancy and lactation also imply that the survival of offspring of a one-parent family is more or less equal to that which would result from a two-parent family, with the male able to contribute very little beyond the protection of mother and young. In consequence, a male's individual reproductive output is best served by multiple mating, which is why, in the majority of mammalian species, males tend to be polygynous, or simply promiscuous.

Implications for human reproductive systems

Inevitably, once we have reached this conclusion, people begin to wonder about the implications for human breeding systems – and it is inescapable that, on the same logic, humans are, evolutionarily-speaking, 'designed' to be promiscuous or, at best, polygamous. Many human societies are indeed actively polygynous and the increase in divorce rates in the western world suggests we are moving rapidly towards a system of serial polygyny. Where the cultural norm is monogamous, it is interesting to note that this is most clearly expressed in extreme climatic conditions (such as colder latitudes) where survival prospects of a two-parent family are significantly higher than those reared by a single parent alone. Even here, however, as also in other primarily monogamous peoples, society has had to develop strong cultural traditions to ensure male fidelity and stop desertion: cultural traditions based at least partly on embarrassment or 'shame' in deviating from what is 'socially acceptable'. I present such comments simply as a factual scientist and make no value judgements, attempting neither to promote monogamy, nor to condone promiscuity. But in a western world where women are vocal in their demands for equal rights, I find myself smiling wryly that subtly and unobtrusively they have already managed what most female mammals cannot achieve: to entrap their men in a cultural mesh of monogamy.

12

Mating behaviour and mate choice

Even now we are only part-way through the story. We may have determined why different species show different forms of breeding system, but we still have not explored what may account for the range of different types of reproductive behaviour shown within that general approach. Once again, the actual behavioural style adopted and the form of the behaviour expressed will be such as will maximise mating success under the particular circumstances. The form of reproductive behaviour in this sense, like that of social organisation or group size, may be fixed for a species, but can also vary from population to population within a species, depending on ecological circumstance.

To illustrate this, let us return again to the model we used in our earlier investigations of the factors affecting group size and explore the remarkable variation in mating systems to be found amongst European deer. Conventional wisdom has in the past suggested for each species a rather specific mating style: such that red deer stags compete amongst themselves to win and defend a harem of mature females; male fallow deer (bucks) 'traditionally' establish an exclusive small mating territory or 'rutting stand' from which they call to attract oestrus females – while roebuck, which during the spring and summer are usually strongly territorial in defence of their entire home range, simply mate with those females whose range overlaps with their own. These conventional stereotypes, however, simply do not stand up to closer inspection and conceal a variation of pattern just as wide as that we described in the last chapter in terms of social organisation.

Let us first explore this variation in strategy amongst fallow deer. As in many species of deer, males and females are separated for much of the year, operating in groups of separate sex and often even occupying discrete geographical areas. Fallow deer are polygynous animals, with each male mating with as many different females as he may encounter, or at least as many as he may win in competition against other males. As we have suggested, conventional wisdom describes for fallow deer that, during the autumn, mature males move into those geographical ranges occupied by females and their followers, where they compete amongst themselves for traditional mating grounds or 'rutting stands'. These traditional mating grounds are widely separated each from the next (by a kilometre or more) and are usually considered exclusive territories held by a single powerful male and actively defended against other males. Successful males, once established on a stand, bellow (or 'groan') to attract oestrus females for mating.

Such rutting stands do exist, but are far from a universal phenomenon. In some populations, males do indeed hold such 'classic' rutting stands, but in others, while mature bucks still do hold exclusive rutting territories – and these are still of much the same size as in the 'traditional' scenario – they are not at any distance from each other but, rather, are clustered in space so that two or three such stands may all abut each other closely in the same area with common boundaries. In a further variation of this basic strategy, the number of males clustered together increases, the size of territory defended becomes vanishingly small – really no more than standing room only – and a cluster of up to 20 or 30 males forms a communal display area in the equivalent to the communal displays of lekking species such as black grouse or sage grouse.

Nor are all variants of the mating strategy based on territorial ownership of land. In some populations of fallow deer, bucks do not establish rutting stands or display grounds at all, or if they do, it is only until they have attracted a likely-looking bunch of females. They then switch their attentions completely to ownership and defence of the females themselves, herding them, travelling with them and defending them against other males in a style more commonly associated with red deer in maintenance of a harem of females. Finally, other males again seem to avoid competition altogether, whether for territory or females, but become wanderers, travelling widely through the females' ranges, mating opportunistically with oestrus females as they come across them.

Actually, from what we have seen so far, such flexibility, rather than rigid adherence to a fixed behavioural strategy, makes intuitive sense, too. Imagine yourself as a fallow buck in an environment of continuous woodland, relatively homogeneous and with a relatively high density of females evenly dispersed throughout that habitat. Here, indeed, the best way of maximising the number of matings achieved (and thus the number of offspring sired, which is after all, what it is all about) may well be to establish an exclusive territory some distance from other males, groan to call groups of females towards you, copulate with those actually in oestrus and then let them move on, making no attempt to herd them or hold them on your stand. After all, like city buses, in an area with a high female density, there'll be another along in a minute.

Now consider your *alter ego* in a different environment – perhaps an agricultural landscape punctuated by a few occasional scattered copses. Here the environment is highly heterogeneous and the dispersion of females may well not be uniform. There may well be clear hotspots where females will tend to congregate to feed, perhaps areas of particularly good grazing. Holding a rutting stand widely distant from other males is no good to you here – if that stand happens to be in an area unfrequented by females. Here is a prime case for moving to the honeypot and establishing your territory there, even if it does mean that it is close to those of other, rival bucks and competition is thus increased.

Finally, indulge your imagination with one further scenario. Consider what might be your mating success on a fixed rutting stand in an environment where females are overall of very low density. If you call females to your stand, mate with those actually in oestrus at the time but then let them move on, your overall mating success may be pretty poor: those may have been the only eligible females for miles! Under such conditions, surely the best option if you want to maximise your mating success is to set up a temporary stand, by all means, but once you have attracted a group of females, abandon your territory and stick with the females themselves. While only one or two might have been in oestrus when you first picked them up, they are all going to become receptive sooner or later, so if you stay with them and defend them against other males, you will eventually get a chance to mate with every member of that harem. And where densities of females are very low indeed, so that they occur as scattered individuals, your best strategy might simply be to abandon any other plans and forage widely through your environment in search of occasional oestrus females.

While such idle speculation makes a degree of flexibility in mating strategy seem eminently sensible, intuition is notoriously unreliable. But in fact, formal analyses by Jochen Langbein and Simon Thirgood of the environmental correlates of different mating strategies in fallow confirm that the mating system alters specifically in response to the density of males (and thus the degree of male-male competition for mating opportunities at all), density of females, and the proportion of woodland or tree cover within the home range.

Although we have focussed our attention for this example on fallow deer, similar variation is apparent in other species too, traditionally regarded as having only one mating system. In some parts of their range, Japanese sika deer appear to be territorial, with mature stags establishing classic rutting stands, defending them aggressively against rival males, often marking trees within the stand by scoring them with the tips of their antlers and whistling to attract oestrus hinds. In other areas, however, sika are described as holding harems of hinds; indeed, in one of the areas where I myself used to work, the general pattern actually seems to have shifted from one to the other over the years. Alongside these, there have always also been occasional 'wanderers' foraging for oestrus females. In red deer, too: a species widely described as harem breeders, some populations have recently been found to establish rutting stands: the complexity continues.

We are not even safe in relying on the stereotype for roe. Seasonally-territorial, mature roebucks establish exclusive territories each spring (of between three and 30 hectares) holding these until the autumn. Males in general hold territories somewhat larger than the home ranges of females – and the accepted wisdom is that males mate with those females whose ranges lie within their own territory. However, recent studies in Sweden, based on very close observation, have shown that such territorial bucks often have another, satellite male associated with them and that females not uncommonly mate with two males (after all, male territorial boundaries do not precisely coincide with the boundaries of female ranges and any one female's range may well fall within the territorial preserves of more than one male). But in addition, studies of the movement patterns of individual females have shown that they do not necessarily mate with the males whose territories overlap their own anyway: a number of does make very purposeful one-day excursions from their range at the peak of oestrus, choosing to go and mate with another buck a number of territories distant.

Courtship displays and mate choice

From what we have been saying in the last chapter, it is clear that much of the complexity of courtship rituals and mating behaviour may be a result of female coyness: trying to engineer that investment by a male should, if possible, be made to approach or even equal her own before 'agreeing' to mate with him – in order that it is in his best interests to assist in any resultant breeding attempt and increase his own commitment to that attempt. In a number of cases, successful courtship involves the bringing of gifts: prior establishment of territory so that one may offer a female her own patch of home ground, even a ritualised courtship feeding where the suitor brings food to the female he wishes to persuade.

In all such cases the complexity of the display required to attract a mate at all vastly increases the male's initial investment, so that it more closely equals his mate's. But the success of such a ploy depends on the male having no option but to play along: the device in practice only works because a female can choose whether or not she will mate with any given male. In effect, our discussions of different mating strategies of a fallow buck, on pages 113–115, considered them from the point of view of a male maximising his own reproductive success; but the female herself is not simply a passive pawn in the mating game and we may now move on to consider things from her perspective.

While it is indeed usually the female who wields the power of choice, it is not always so, and we are in fact back once again in the realms of greater or lesser investors. More accurately, we must concern ourselves again with questions of whether or not either sex's reproductive success is more limited by availability of resources or availability of mates. If one member of a breeding pair has to invest heavily in each reproductive attempt and can thus afford very few, she must ensure maximal success of each breeding attempt on which she embarks and should be very choosy about the partner she selects, in terms of his general fitness and genetic quality, as well as in the likelihood that he will stay with her and assist with the protection and rearing of any young. The reproductive success of that sex which does not invest heavily in terms of resources in this way is, by contrast, mate-limited. The resource-limited female (in most cases, of course, it is the female) has him over a barrel: she must be choosy to protect her investment – and he wants to be chosen, because the whole basis of his reproductive success is mate-limited.

Thus there is an important further consequence of any disparity in reproductive investment: that the greater investor may express power of choice in terms of mates. And it is precisely because she can choose that the female is able to 'force' upon the male the requirements for further investment, to try to engineer that his investment in any breeding attempt comes to equal her own and that it is therefore in his best interests, too, to stay with her and co-operate in the protection and care of the young she may produce – to move him from promiscuity towards monogamy.

But the other side of the coin is that there is an onus on the female (or on whichever sex is resource-limited and thus the chooser) to ensure she chooses the best mate for each reproductive attempt in which she may invest, maximising the quality of her offspring and their chances of survival. On what basis should she choose, and what clues are available to her to guide that choice?

Mechanisms of choice: courtship displays

Clearly from what we have just said, courtship displays are a primary mechanism by which competing males may advertise their wares to choosy females. Courtship displays in fact serve a variety of purposes: species-specific behaviours, calls or colouration ensure selection of a correct mate from among your own species. Long-distance advertisement (such as the carrying calls of groaning fallow buck) may serve to attract mates from a distance and ensure that a suitable pairing coincides at the same place and at the same time. There is an increasing body of evidence to suggest that the stimuli derived from initial courtship may have a direct effect on hormonal levels of both partners, bringing them into full reproductive condition – and into full hormonal synchrony to permit a successful mating.

The same displays commonly also serve to suppress for a period the normal tendency towards aggression between two individuals for long enough to permit mating to occur at all. Where a pair may stay together beyond simple mating, perhaps to co-operate in the care and rearing of the young, courtship displays may be prolonged, continuing over a long period to maintain this suppression of aggression, cement the pair-bond and maintain hormonal synchrony throughout the breeding period. The demand for prolonged courtship displays, as we have suggested, may be a device whereby a female may force a potential mate to increase his investment to more closely match her own; but perhaps the major pressure behind such display is the need to be chosen. That sex which we define as mate-limited must advertise its qualities to seek copulations; the sex which invests so heavily in any one breeding attempt that it can make only a few such attempts must choose the most suitable consort.

What are the qualities which should influence choice?

There are a number of qualities on which the chooser should select a mate. Remember, she (let us simplify things by assuming it is the female) must choose a male whose contribution to the breeding attempt will maximise the survival of her offspring. This means, in the first instance, that he should have 'good genes', making him a survivor, and heritable, so that those same 'good genes' are passed to his offspring improving their intrinsic fitness. She cannot judge that directly, but the quality of a partner's genetic make-up should be, at least in part, reflected in the expression of those genes in the phenotype, so she may judge by appearances, or apparent quality. But intrinsic fitness on its own is not all she may select; amongst those species where the reproduction involves more than the simple provision of sperm and egg, and in which one or both parents may invest in their offspring with some degree of parental

Male European kingfishers contribute to the breeding attempt by bringing food back to the nest

care, the chooser should select a mate who is likely to assist with the protection and rearing of her young.

Finally, good genes on their own are not enough. The whole basis of natural selection is that there is significant variation within any species in genetic makeup; further, many characters are coded for by complicated 'sets' of genes acting together. To ensure that her own successful genes are strongly represented in the genetic constitution of her offspring, and that successful 'gene sets' are not broken up or countermanded by the genes provided by her mate, the chooser should seek a mate whose genes are not only of quality, but which are compatible with her own.

What evidence can she seek to inform her decision and how can she be sure she is making the correct choice? After all, in the same way that she is not a passive partner in all of this, neither is the male. And it is in his interests to be chosen (by as many females as possible) so that selection will have encouraged him to emphasise those qualities that females seek to judge. The whole panoply of secondary sexual characters – bright plumage, striking colouration, loud calls, acrobatic displays – have all developed in response to the fact that females are choosy and the characteristics emphasised in display are, of course, the ones which evolutionary experience has shown are those by which a female may judge. The importance of being chosen is so pressing that these secondary sexual characters are developed to the hilt to try and convince the choosy female of his unparalleled quality. And he may try to cheat.

Judging fitness

Since in most cases a male's reproductive success is limited for the most part simply by availability of partners, males compete strongly amongst themselves to secure mates. In highly polygynous mating systems – in which reproductive success of individuals is strongly skewed so that successful males may mate with many females, while others, less successful, may get no matings at all – competition is exacerbated and may become intense.

In such cases, a choosing female may base her assessment of male quality to some extent on the basis of the outcome of male-male competition: winners of disputes over mating title are clearly stronger and thus likely to be fitter mates than losers. Amongst species which defend breeding territories, males capable of securing territory, where resources are perhaps limited, are clearly fitter than those which fail to gain a territory. Even amongst territory holders, further refinement in choice may be based on the relative quality of the territory occupied. But commonly, contests between males are simply a matter of display; in other cases, any competition between males is indirect: they display not to each other, but to watching females, with winners and losers amongst the males themselves simply determined by the results of female choice.

What cues may a female use in such situations to judge quality? Commonly it appears that females select on the basis of characteristics which in some way are themselves indirectly determined by underlying physical fitness. The duration and intensity of a male songbird's song, for example, must ultimately be limited by its general fitness: the amount of 'puff' and the spare resources of energy and time it can afford to devote to singing (rather than needing to devote every waking moment to searching for food because he's really up against it). Given equal foraging opportunities, the simple size of the body may be a good indicator of thriftiness – or a sign of dominance in more social species, revealing an ability to sequester the prime food resources and displace weaker individuals. Mating success in red deer, for example, is closely related to body mass – and the size of antlers – because a male who can 'spare' from his daily foraging, the energy and mineral requirements for the development of the finest spread of antlers is clearly relatively fit. In fact, to be more accurate, body mass and antler size in this case determine the outcome of male-male competition for mates, while females actually select males on the basis of their mating call or 'roar'. But the quality and timbre of that roar is itself determined by body mass (with a deeper roar emanating from a stag of deeper chest) and, as before, the frequency and duration of roaring accurately reflect the amount of time and energy that individual has 'spare' to invest in self-advertisement.

Most of these rather general features are directly correlated with male fitness, but the same principle may be extended to the development of secondary sexual characteristics or 'ornaments' used specifically for display. The bright colouration of many male birds, for example, while in part designed as a signal to cue correct species-recognition (page 55) may be exaggerated as a demonstration of fitness or dominance.

The width of the central black stripe down the chest of a great tit, for instance, is closely correlated with dominance, rank and status within the male population, so that a female choosing a male with a broader stripe is accurately selecting a male who has thus far proved to be one of the more successful amongst his peers. And the very fact that female choice may be based on the quality of such ornaments – and that mating success is thus determined by the quality of ornamentation – acts as a powerful selection pressure on males to enhance the quality of the ornaments by which females choose.

This may, in effect, lead to them becoming exaggerated to such an extent that they are not longer accurate reflections of actual fitness – beyond the fact that a male who can afford to devote so much of his limited energy resources to development of such exaggerated ornaments must have plenty to spare and thus be generally rather fit. But the continued reliance by choosy females on the 'quality' of these same ornaments may lead to such pressure for their exaggeration that, in the limit, they become actively disadvantageous to the male: bright colouration making him more apparent to predators, or an exaggerated tail hindering his ability even to fly.

The classic extreme example here is in the evolution of the peacock's tail. Females do indeed select males on the basis of the size of the tail, the number of 'eyes' and the frequency with which the tail is shimmered in display. But to the male, while the tail is obviously of crucial importance as a display feature with which to secure mates, it is otherwise actually an encumbrance. It is energetically expensive to produce and makes flight rather difficult. It might be argued in effect that sexual selection promoting the development of the most elaborate tails has now gone so far that females should presumably select males with less extreme appendages, since these are probably actually rather fitter.

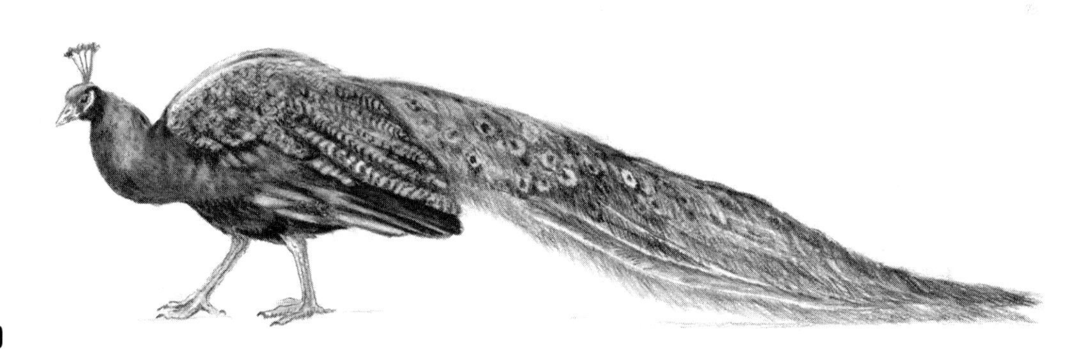

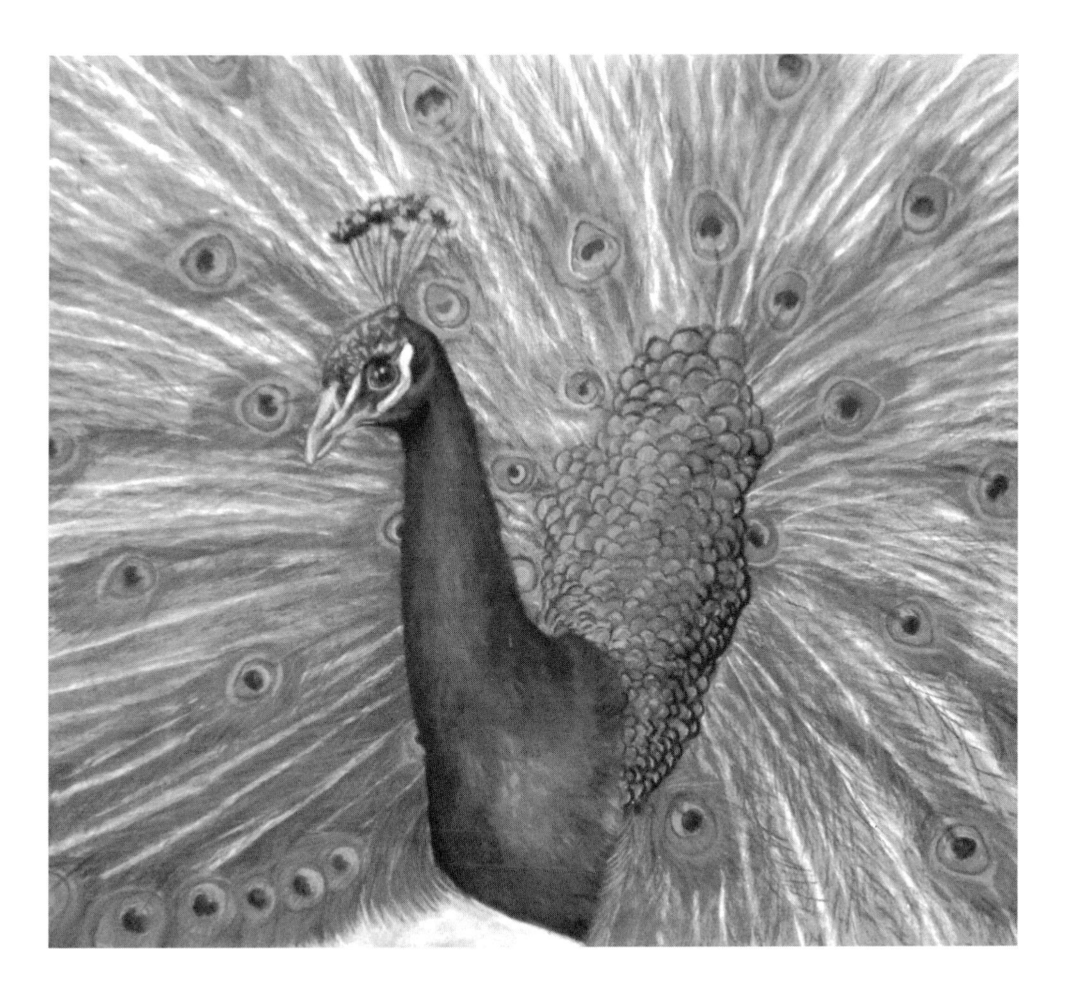

Two theories have been put forward to explain why females should still prefer males with longer and more elaborate tails, despite their apparent disadvantages. Firstly, it is suggested that males with the most elaborate tails must be of superior quality, since they are still successful, despite their obvious handicap. That they can survive and have sufficient resources left to produce these extravagant ornaments, despite the handicap they must represent, surely assures that they are truly fit?

An alternative explanation makes the point that the whole essence of natural selection is that animals should seek to maximise not simply the number of offspring they themselves produce, but their genetic contribution for generations to come; that, in short, one should not 'seek' merely to produce the maximum number of one's own offspring, but that these offspring themselves should achieve good reproductive success and that thus one's own particular genes come to represent an ever increasing proportion of future generations. In that case: if most females are still selecting mates on the basis of a long tail, successful reproduction of your own sons can only be assured if they, too, have long tails.In such a circumstance, whatever the handicaps and however decoupled the signal may be from actual fitness, you should choose the most extremely ornamented male to ensure that you yourself have sexy sons.

Judging a mate's ability to provide

Another way, of course, in which a female may judge a male's fitness is by his 'possessions'. But in this case too, she is also able to judge his ability to provide for her, and his commitment to assisting in the protection and rearing of any offspring. This also, as we have seen, may be a significant factor in her choice of a mate – since if he is merely a provider of good genes but makes no further effort to ensure survival of the offspring, he may not, after all, be the ideal choice.

As we have noted, in many species courtship involves a good deal more than simply showing off one's own finery. Males of many species must first establish themselves on breeding territories before they may attract a mate. The winning of that territory against the competition may be an immediate reflection of fitness, but ownership of that territory and the quality of the resources it contains are also a reflection of the male's ability to provide for the female and, in due course, her young. In many species, too, courtship displays have become extended and actually incorporate a period of courtship feeding in which the male brings food to the female as an integral part of his display. While sometimes the food-giving is representational only, in other cases the gift is real (and the added intake may even be essential to success, in providing the female with the extra energy required for the development and production of a clutch of eggs). In either case, the 'willingness' of a male to provide nuptial gifts again reflects something both of his intrinsic fitness and his likely abilities as a provider after mating.

Genetic compatibility

Good genes may not be enough on their own. As we have discussed, selection of a mate may in some measure depend also on some estimate of the compatibility of his genes with your own. It is fairly generally-accepted that there are tremendous disadvantages in breeding with individuals too closely related to oneself. Such inbreeding may lead to the development of genetic disorders, or to the emergence of recessive genes which are potentially damaging (but whose expression is normally masked in combination with more dominant genes).

Less widely recognised is the fact that there are also disadvantages in breeding with individuals too dissimilar to oneself. In such a mating, particular combinations of genes in one individual which have perhaps been developed over generations by repeated selection may be broken, or masked, by the genetic input from a totally unrelated individual which has adapted its genes along some different route. The offspring may fall between two stools and gain the fitness of neither parent. Considerations of optimal outbreeding actually suggest that to avoid the pitfalls of inbreeding depression and to avoid risks of loss of one's own genetic 'pattern', the most effective genetic choice in such regard is to mate with one's first cousin.

Such a choice would require remarkable discrimination, for while human beings may be introduced in time to their genetic relatives, how would animals come to recognise their cousins or, more accurately, how could they distinguish the degree of relatedness to themselves of any new acquaintance? Actually, it is not so very difficult to see how this level of outbreeding could be achieved on average, whether an individual may recognise its relatives or not. Given the actual dynamics of most animal populations, and the fact that individuals disperse a limited distance from the place in which they were born, a series of concentric rings drawn around one's birthplace fairly accurately predicts the degree of relatedness of other individuals to yourself; the further you travel from your home, the less related to you will be other individuals of your species. In search of optimal outbreeding opportunities, then, in terms of degree of genetic similarity/dissimilarity to yourself, you only need to travel a given distance from your home.

And it would appear that that is indeed what many animals do: dispersing from their birthplace, when they reach maturity, to just such a distance which would ensure optimal outbreeding. Within that zone of 'optimal-relatedness', where almost every individual will be appropriately related to yourself, then selection of a mate may be based on actual fitness and ability as a provider as already discussed, in certain knowledge of suitable genetic relatedness amongst potential mates.

But we should not imagine that animals are not capable of distinguishing differing degrees of relatedness to themselves even amongst individuals they have never previously encountered (so that the judgement must be made on physical characteristics rather than on prior knowledge). In a delightful set of experiments, Patrick Bateson reared broods of Japanese quail in relative isolation from each other. When they reached maturity, he offered individual females a choice of potential mates, giving them the choice between brothers, first cousins, second cousins and animals only very distantly related to themselves. The majority of females selected first cousins, even though they had no prior knowledge of those individuals. How did they choose? In Chapter 7 we discussed how in that critical period of social learning we call imprinting, animals learn the characteristics of their mother, and then can generalise from that particular to the more general case to give themselves a 'model' on which they may base future recognition of their own kind. Perhaps the mechanism of generalisation is such that one does not simply generalise from 'mother' to 'any female' but can actually generalise in different 'bands' to be able to discriminate greater and greater levels of dissimilarity? Bateson repeated his experiments, but this time he switched the broods, so that each female destined to be the subject of future experiments was actually cross-fostered and reared amongst her first cousins. Tested again at maturity, females again selected mates very carefully, but in this case did not choose their own first cousins as mates, but those which were first cousins to her foster-siblings (first cousins of her own first cousins). Clearly each female could accurately distinguish degrees of relatedness, and was indeed developing the 'model' from her early experience of (foster-) mother and of littermates.

But whatever the model, our point here is this: that even from simple physical characteristics and subtle differences between individuals previously completely unknown to them, Japanese quail can readily discriminate precise levels of relatedness to themselves with no prior knowledge (and without dependence on other correlates like distance from home). Similar studies have since been done with other species and however surprising it may seem to us, this ability to distinguish subtle differences in the level of relatedness of unknown individuals would appear to be fairly widespread. We should never underestimate what we cannot emulate.

How to compare? How to choose?

Our human ability to underestimate the mental capacities of other animals is also a potential challenge in understanding how females actually choose between the diversity of potential mates on offer. Where groups of males come together and display on some communal display ground or lek – like blackcock or (exceptionally) fallow deer – then the decision may be relatively more simple: all the candidates are present together and the female may make a direct comparison and a direct choice.

But such opportunities for simultaneous comparison are relatively rare in the animal kingdom and more commonly a female will encounter a number of different males, one after another, over a period of time. How can she possibly keep an accurate memory of all their various qualities in her head and ultimately return to her choice?

Clearly some animals can select a mate from such sequential comparisons: the enterprising female roe deer of page 116 opted for an excursion from her normal range (and the immediately obvious male whose territory overlaps her own) to visit another male far from her normal range, presumably on some memory of his quality from a previous encounter. But simpler devices are in fact available for those whose memories may be poorer! Females may simply mate at random, although this implies no selection whatsoever. Alternatively, they may employ some fixed-threshold method of discrimination, by which they inspect males sequentially, judging them in turn against some fixed standard and mating with the first male they meet whose quality exceeds their minimum threshold.

In a more complex strategy, females may inspect males sequentially in just the same way, but continue until the one most recently encountered is of lower quality than the one before; at this point she stops searching and returns to the penultimate male. In this case, in effect, she compares each male with the one encountered previously; if he is an improvement she rejects the one before and moves on, continuing her search as each new male proves better than the last, until she reaches the peak and takes a down-turn. Although there may be better males yet she has not encountered, this method at least ensures that she accepts the best male in her sample so far - with minimum taxing of memory. Finally, a female may adopt what is known as the 'best of n' strategy: genuinely sampling all males she can and retaining some memory of the qualities of each. When she runs out of males, or time, she can return to the best male encountered during her period of search.

Clearly these last two methods assure a somewhat better final choice than does random mating or a fixed-threshold strategy – but even so, a female is still limited to selection of the best male within a limited sample. There are considerable costs in sampling: by abandoning one male for the time being and moving on, she may lose him to another female and find him already mated when she finally decides he was the best option and returns; by sampling too extensively, she may pass beyond a limited period of receptiveness (such as mammalian oestrus) and so on. Further, given limited time to judge each male's quality, how accurate may be her assessment anyway?

A number of behaviourists have argued that, at least amongst polygynous or promiscuous species, where males are not limited to mating with a single female, it may prove the best strategy for any female simply to copy the decisions of others. Copying the choices of others in this way may be viewed as a form of hedging your bets: rather than risking a bad choice, why not copy and, at least, do no worse than the female being copied? In effect, copiers are adding to any assessment they may themselves make of a potential mate's fitness, elements derived from assessment of its previous mating success. In fact, copiers should at least do no worse than choosers, which ignore mating success and mate according to their own personal assessment, because they are 'integrating' the experience and assessment of other females beyond themselves. And where assessment is difficult and quality hard to judge, the average fitness-choice of copiers may be higher than that of individuals who try and make their own choice. Such conclusions may help explain the extreme skew in mating success observed in many lek-breeding animals, where one or a few males gain the vast majority of all matings.

Cuckoldry and extra-pair copulations

However thoroughly we may try to analyse the underlying logic of animals' mating systems and however compelling our arguments, the animals themselves are always one step ahead! Over the years – and especially since science offered a way of undertaking paternity tests through analysis of parent and offspring DNA – it has become more and more apparent that even amongst animals and birds whose strategy is generally recognised as monogamy, a surprising proportion of offspring are fathered outside that supposedly faithful pair-bond. We have already argued that amongst birds in particular, where the amount invested in any given reproductive attempt may be more or less equally distributed between the parents, we might expect that the optimal reproductive strategy for both sexes – in terms of maximising their own individual reproductive success – would be to form stable monogamous pairs (page 110).

Yet in a surprising number of cases and in a surprising number of species it turns out that in any given brood, one or more offspring will be the result of a copulation outside that mated pair. Thus for the common European blue tit it is estimated that 36% of broods show signs of extra-pair paternity, with 18% of offspring actually sired by extra-pair matings; for red-winged blackbirds the proportions are even higher, at 47% and 28% respectively.

Is this just chance promiscuity? More probably we may consider it as hedging one's bets? Or the concept of copying others' choices somehow inserted into an otherwise monogamous mating system. Within such a system, while a female may honestly have tried to assess the quality of a potential mate, she can never be 100% confident that her assessment was truly accurate. Thus extra-pair copulations, copying the choice of another female and mixing the paternity of a given brood of youngsters, may give some guarantee of fitness to at least some of her offspring, if her own initial choice was incorrect. Or maybe, despite her choosiness, it is simply clear that the male next door is, in any case, a superior specimen than her own mate – so that if she can solicit an extra-pair copulation, she may have the benefits of a fitter progenitor for the offspring she rears within her 'proper' pairing. But this is a risky strategy, because it is likely that she may receive reduced support in the breeding attempt from a mate who becomes aware of being cuckolded: males should actually invest less in broods for which they are not confident of paternity.

Whatever the evolutionary explanation, it is clear that such cuckoldry is far from uncommon in the animal world; indeed, it is even argued that as many as 4% of human children worldwide were fathered by someone other than the man who believes he is their true biological parent.

13

Co-operative breeding and arguments about altruism

While we have been at some pains to establish that natural selection acts at the level of the individual – and indeed acts through each individual attempting to maximise the number of copies of itself it may leave to future generations – there are occasions when strict adherence to this rule appears to falter, when individuals appear to suppress their own reproductive activities while at the same time assisting other individuals to increase their reproductive success. And indeed, more generally, there are situations where any one individual would appear to be behaving in a way which actually reduces its own 'evolutionary fitness', perhaps reducing its own reproductive potential, or increasing its own risk of death or injury, while offering advantage to others. Such behaviours appear to us to be altruistic.

But how could such behaviours evolve? If they have a genetic basis, such that there is a gene controlling a specific type of altruism, then if the consequence of that altruistic act is to reduce the number of offspring left to successive generations by any individual who possesses that gene, while possibly increasing the number of offspring left by those it helps, the gene controlling altruism will inevitably become rarer and rarer within the population simply because those who show it leave fewer offspring than those which do not. The only way in which early theorists could account for this was to suggest that perhaps selection does not only act at the level of the individual, but on occasion may act at the level of the social group, the population or even the species, over ruling considerations of individual advantage. Proponents of such theories of group selection considered that, in certain situations, an animal might sacrifice its own selfish, individual advantage, for the good of the group, or for the good of the species.

In general, of course, actions or behaviours which were for the good of the species as a whole would also be to the individual's advantage and so the two levels of selection would act in the same direction; on occasion, however, the interest of individual and group might be in opposition, resulting in a tension between the two which might lead to the appearance of behaviours not strictly to the individual's best advantage. There is, in fact, rather little evidence for any form of selection at any higher level than the individual (and, at least within the limits of our current understanding of genetics, no real mechanism which would allow evolutionary change against the individual's own advantage). We can if we wish, however, leave the case open: as we have just noted, selection pressures at the level of the group would be, in the majority of cases, in exactly the same direction as those acting on the individual, so we cannot actually disprove some degree of selection acting at the higher level. But the point is that, in practice, there is no need to invoke such a level of selection. All the various examples we might come up with which appear to constitute some form of altruistic behaviour can in fact be shown not to be altruistic at all, or in any case can be explained by existing theories of selfish natural selection.

Distraction displays

Many animal species show specific defensive behaviours in protection of their offspring: attacking or 'mobbing' potential predators or performing some attention-grabbing behaviour to distract the predator's attention away from a nest or mobile young and onto themselves. American killdeer and European species of plovers, as well as many other wading birds, have a specific 'broken wing' display in which the adult flutters, apparently helplessly, on the ground, but always just out of reach of the predator as she draws it further and further away from nest or chicks.

But this apparent altruism is, of course, nothing of the sort. The essence of even individual, selfish selection, is that each individual shall leave as many surviving offspring as possible to future generations and protection of the young from potential predators is merely part of that selfish assurance that as many of the offspring as possible do indeed survive to continue the line.

Co-operative breeding

Amongst some other species, however, there is another whole suite of behaviours associated with reproduction which are not quite so easily explained away. In a number of species of both birds and mammals, some individuals within a social group or a population do not themselves breed (even though they have reached adulthood) but actually assist in the reproductive activities of others as 'helpers' or 'aunts'.

Amongst mammals this is particularly commonplace in the dog family. Within social groups of dingoes, wolves or African hunting dogs, for example, sexual activity is often restricted to a single dominant female within the pack (in the same sort of reproductive suppression noted amongst social insects on page 102) and other females, while they have reached adulthood, are sexually inactive. These subordinate females however often assist with care of the young and may regurgitate food for them when returning to the den after a hunt. This 'helping' behaviour is also found in other social carnivores like mongooses or meerkats.

If the summer is good and there is a plentiful supply of insect prey, European swallows or house martins commonly attempt to cram two or (exceptionally) three broods into a single year. Subadults from the first brood are regularly found to help at the nest, bringing food and sharing with the parents the care and feeding of subsequent broods. These birds are relatives and, as we shall see, there may be a special premium in assisting with the care of your immediate kin. But there are also examples in the bird world of assistance at the nest of breeding pairs from birds who are not related to them at all (as, for example, among the green wood hoopoe, the Florida scrub jay or acorn woodpecker). Here are other adults, presumably perfectly capable of breeding in their own right, apparently forgoing their own opportunity while assisting the breeding activities of other unrelated adults. Why should they behave like this?

In fact, there are a number of factors at play here. In the first instance, it is not at all clear that in many such cases the apparently altruistic helpers are in fact giving up the opportunity to breed themselves. Subadult housemartins assisting with the rearing of later brothers and sisters are themselves not yet reproductively mature. Subordinate female wolves or dingoes have actually been effectively neutered, in that the strong dominance of the breeding female suppresses completely their own production of reproductive hormones. And the incidence of nest-helping among wood hoopoes is highest in populations at high density. Wood hoopoes are territorial; at high population density, the environment may quickly become saturated and it is quite probable that many helpers are actually unable to establish territories in their own right at all.

While this might explain that there is no reproductive loss involved for themselves, it nonetheless does not explain why they should bother to help the reproductive attempts of

In dingoes, as with other wild dogs, breeding is restricted to the most dominant female. Other females, while they have reached adulthood, are sexually inactive

others. But, once again, there may be a number of selfish advantages. Amongst both birds and mammals there is very good evidence to suggest that reproductive success improves with experience: that the survival of a second litter or brood is significantly higher than that of the first breeding attempt of an inexperienced animal. If you cannot commence a breeding attempt yourself anyway, then helping others while still relatively young yourself may enable you to gain valuable experience, so that when you are subsequently able to breed, you have maximum success on the first attempt. And research data show that many of those animals which have previously been 'helpers' do indeed have higher breeding success, when they themselves start to breed, than those which are completely naive.

Helping others may also be of advantage in the longer term, even in the sense of helping you get started at all. Nest-helping wood hoopoes gain familiarity with the territory of those they help – and have a statistically higher chance than do complete outsiders of 'inheriting' or taking over that territory themselves when the current owners die. Subordinate wolf bitches who assist with the rearing of another's pups have a higher chance of rising in due course to become the alpha female in their turn.

Kin selection and inclusive fitness

Finally, as we have noted, while nest-helping wood hoopoes are not necessarily related in any way to the breeding pair whom they are assisting, in many cases of breeding cooperation, helpers are close relatives of those who are receiving help. And there is a logic which says that, since those you are helping have many of the same genes as you do yourself, then in assisting them with their reproductive efforts – given that you cannot reproduce yet in your own right – you are still increasing the number of your own genes passing on to subsequent generations. While natural selection actually acts inevitably on the living organism (and thus the expression of its genes, or the phenotype), the actual genetic material involved in inheritance of advantageous characters is, of course, contained in the genes themselves.

In fact, because production of offspring involves two parents (and offspring inherit half their genes from each), your own progeny contain on average only one half of your genes (and one half of your partner's). Brothers and sisters of the same mating each also have half of the parent genes, so that your brothers and sisters again on average share half their genes with you; their offspring, inheriting half their genes, have on average a quarter of your genes, and so on.

If selection acts to promote the proliferation of advantageous genes, then the survival of two nephews or nieces is as advantageous to the individual as the survival of one of its own offspring. This all sounds rather dry and statistical, but is nonetheless a valid logic. In terms of passing on your own advantageous genes, survival of the offspring of close relatives adds to your own inclusive fitness, and investment in those offspring may be as effective as having offspring of your own - or at least may add to the genetic legacy you can achieve through your own reproductive efforts. To bring it closer to home: while I may choose to have children of my own, these would each pass on only a half of my own genes. Between them, my sisters have a total of seven children: each of these nieces and nephews actually shares one-quarter of my own genes. Assisting my sisters with the rearing of only two of those nieces and nephews (while costing far less than having children of my own) is actually the genetic equivalent of my producing a single child myself.

Green wood hoopoes at nest

While such arguments might appear to be taking selection beyond the individual, something we argued could not occur in practice, in fact they are only extending it as far as a close kin-group, and the close genetic relatedness in such cases – and the fact that selection acts to promote the transfer of advantageous genes, however that may be achieved – means that investment in kin in this way actually does confer an advantage on the individual in terms of its inclusive fitness, or the overall transmission of its genes to future generations.

Purely mathematically, a gene which controls helping behaviour will spread rapidly through such a kin group since, by increasing the survival of its relative's offspring, which themselves are likely to possess the same gene, an individual's helping behaviour in itself increases the frequency of the 'helping gene' within the local population.

Alarm calls

One of the other classic examples which is often cited as a challenge to individual selection is that of the alarm calls uttered by many social mammals or birds if a predator is noticed nearby, calls which 'warn' other members of the group or feeding assemblage of potential danger and allow them to take some evasive action. Many small songbirds react to the presence of a predator by such calls, as do ground squirrels, marmots and other small ground-living rodents. A suspicious European rabbit drums with its hind feet on the ground and flashes its white scut of a tail as it runs away. European roe deer also fluff up the hair of their white rump patch into an extremely obvious 'powder-puff' if they scent or see a predator and run off, 'barking' loudly. And perhaps the most extreme display among the antelope is that of the African springbok who flare their magnificent white rump patch and make a series of high leaps into the air, 'pronking' ostentatiously as they run (leaps which surely delay their own departure). Surely, one of the best ways of avoiding predation yourself is to avoid detection, whereas actually calling or other behaviour which advertises the presence of the predator to others must act, quite by reverse, to draw attention to yourself, while allowing others to escape.

The utterance of alarm calls (or performance of other warning behaviours) in this way would appear to be a clear case of reducing your own fitness (and survival chances) in warning others. We might explain it in some instances, amongst tight social groups of closely related individuals, by invoking the same arguments of kin selection we have just discussed but, in fact, such warning behaviours are relatively widespread and occur just as commonly in large herds or feeding flocks of completely unrelated individuals. In fact, the loss of fitness in such advertisement may not be as great as it appears – and the behaviour may, on closer examination, actually turn out once again to be being entirely selfish. The caller has already the advantage that it has spotted the predator or other source of danger. The call itself alerts others of its group to the presence of the predator, but gives no information about its location – so that in effect it may panic the rest of the flock or herd, leading them to mill about, uncertain as to where (or what) the predator may be and therefore which might be the best direction for flight. This makes them very obvious to the predator (who may not otherwise have noticed others in the group apart from the caller, and thus would have focused all its efforts on catching that one individual) and at the same time very vulnerable. Against this smokescreen, the caller itself may sneak away to safety. In many cases, callers actually remove themselves to comparative safety first, before calling at all to alert others in the flock – while the alarm calls of many small birds are in any case of a very distinctive character and tone, which makes them almost ventriloquial and, at best, very hard to locate. In practice, then, the caller may not be attracting especial attention to itself (particularly if the call is hard to locate in this way) but rather ensuring its own clear get-away amongst the confusion of its fellows.

For many predators, in addition, successful capture of prey is dependent on the element of surprise. A calling songbird knows exactly what and where the predator is, as does the fleeing roebuck or the springbok. And perhaps their display, in drawing attention to themselves, is saying to the predator: 'Look, I've seen you; you've lost the element of surprise and I am not worth pursuing', while the pronking of the springbok (or similar 'stotting' observed in some other species of gazelle) adds to the message: 'And, what's more, I am extremely fit and capable of these high leaps, so you'd be better off finding another, weaker or sicker individual to chase than waste your time on me!' Thus in effect, the alarm behaviour we observe is directed more as a signal to the predator than as a warning to the signaller's colleagues (although, once again, it may gain some selfish advantage from their initial confusion). Most behaviourists now consider that the giving of alarm signals confers more advantage on the signaller than it incurs cost. And as long as the costs are low, there is still a possibility that the behaviour might evolve: because as long as the majority of individuals within a group will signal the presence of a predator, then, if in doing so yourself there is minimum cost to you and others may benefit, there is always the chance that at some time in the future you will yourself benefit from some-one else's call.

Reciprocal altruism

This last suggestion introduces one other possible selfish justification for apparently 'altruistic' behaviour: the probability (as long as the behaviour is commonplace within the population) that some day you may yourself benefit from the 'altruism' of others. Alarm calls and warning signals are very good candidates for explanation by this concept, and indeed we may explain examples of active group-defence against predators – like the defensive rings of buffalo or musk-oxen, or the 'attack parties' of male baboons) in much the same way (page 88)

The requirements for the development of a behaviour by this mechanism of reciprocal altruism is that the cost of the 'altruistic behaviour' must be low (such that actually expressing that behaviour does not greatly reduce fitness of the performer) and that there should be a reasonably high probability of the performer becoming the beneficiary in due course of similar behaviour from some other individual. (It doesn't have to be the same individual as the one he/she initially helped). As long as the behaviour is itself of low fitness cost to the performer; as long as the majority of animals in the population themselves show that behaviour and as long as the circumstances where help is needed are common enough that the initial helper is likely to benefit from help in due course in return, such a behaviour should evolve – through simple individual advantage in anticipation of future help oneself.

Cheating

Apart from the fact that the initial costs to the helper must be relatively low and not life-threatening, the essential requirement for the development of reciprocal altruism is that the majority of other members of the population of society must also show helping behaviour when necessary. Otherwise the probability of other individuals providing future help remains small, and the costs to the initial 'altruist' stand little chance of being repaid; in such a situation, the behaviour will not persist, since costs (however small) are less than the average returns which may be anticipated.

We can explore this further by considering what would happen in such society to a cheat. Let us suppose that there is some relatively simple genetic basis for helping, but that there is a mutant gene which results in cheating: accepting assistance from others around you, but never reciprocating. Clearly, in a society where mutual or reciprocal help is the norm (and thus the gene for helping is common) such a cheat (if nothing otherwise distinguishes it from 'normal' members of the population) is initially at a huge advantage. It benefits from help whenever required, without incurring the costs of ever having to help in return. As long as it may remain undetected (a kind of 'cryptic' cheat) it has a large fitness benefit over 'helpers' and thus will be successful in leaving a good number of copies of itself to future generations. The gene for cheating will increase in the population.

But now let us consider the position a few generations on: the gene for cheating was successful and has increased in frequency within the population to the point where it is as common, or more common, than the gene for helping. Now cheating does not confer the same advantage as before. Because the majority of individuals around you are also non-helpers, the chance of receiving assistance yourself when in need has declined and you are no longer able to get something for nothing. The average fitness of the population is now actually less than what it was when the majority of individuals were helpers; there is also likely to be a strong selection pressure for mechanisms of identifying cheats before offering them help, so that a new advantage arises amongst a small subset of the community who do help each other but recognise and exclude non-helpers.

In short, cheating in an altruistic society is only of advantage when the frequency of cheats is very low. This re-introduces a rather more general idea that we first considered in Chapter 9: that in effect the utility, or advantage, of any behaviour is rarely absolute, but depends to a large degree on what others around you are doing. That is, behaviours that we observe are quite commonly not immediately what we would predict would be optimal for the individual in isolation - purely because that individual does not exist in isolation, and what is optimal in practice will depend heavily upon what other individuals around it may be doing.

Considering, above, the position of the cheat in a population of reciprocal altruists, we find that the relative advantage of helping or cheating depends on the frequency of helpers and cheats within the population. Whatever strategy might appear to confer optimal advantage to the individual in isolation, we may recognise here again that the actual strategy which will become predominant within a population or social group will actually be "that strategy which if adopted by the majority of individuals within a population, cannot be bettered by any alternative strategy" – which we have already described as an Evolutionarily Stable Strategy. We will revisit this idea again, when we talk about agonistic or aggressive behaviours in Chapter 14.

Altruism in human societies ?

Behaviourists, and evolutionary biologists in general, tend to develop a certain cynicism about altruistic behaviours and commonly extend that same cynicism to human societies. Certainly it would seem to be fair to suggest that any examples of apparent altruism which might suggest themselves can usually, with a little further thought, be discounted or dismissed in terms of kin-theory or as examples of reciprocal altruism. But the fact that one can account for them in such away in terms of purely individual, selfish, selection does not mean that one must. Just because one can account for all instances of apparent altruism on purely individualistic grounds does not inevitably mean that one should necessarily do so – and I must leave the reader to make up their own mind as to whether altruism does (or could) exist in human society.

14

Territoriality and aggression

While much of what we have been discussing in the last few chapters, in terms of social organisation and reproductive behaviours, involves co-operation and at least some level of sociality: in and amongst those discussions we have already introduced one or two classes of anti-social behaviour. And indeed such antisocial, or antagonistic behaviour towards other individuals is just as much a part of the social order as active co-operation.

Inter-individual aggression emerges largely in response to competition for some resource – whether it be food, mates or nest sites – and serves either to drive away another individual, to subdue it, or at least to prevent it from interfering. Since we have established that natural selection emphatically acts at the level of the individual, so that each must try to maximise its own reproductive legacy to succeeding generations, it is inevitable that individuals will frequently find themselves in direct competition in this way. In fact, some form of aggressive behavioural interaction is perhaps more to be expected than active sociality and co-operation.

Aggressive behaviour, or behaviour antagonistic to other individuals, may be expressed in a number of different ways: through the establishment of exclusive territories by complete individualists, in the maintenance of individual distance, or a ring of 'free space' around individuals in a loose aggregation – a herd of wildebeest or a foraging flock of sparrows, or finally in the development of a dominance hierarchy, or 'pecking order' of rank and superiority amongst the individuals of a more persistent and stable social group.

Territoriality

The establishment of a territory implies 'ownership' and defence of a geographically-defined patch of ground, within which the owner has undisputed access to the resources contained – whether that be food or mates. But while territories are often held by individual animals or breeding pairs, they may also in some instances be established and defended by social groups – as in the European badger, where a social clan of more or less closely-related individuals may defend a communal territory. In addition, territories are not in all cases necessarily fixed in space: some animals defend mobile territories as a defined area around some equally mobile resource (as an insectivorous bird might defend a swarm of insects as it moves around during the course of a day, or a territorial male Speckled Wood butterfly defends a patch of sunlight on the forest floor and moves with that sunspot, as it moves with the movement of the sun itself).

Territories may be large and contain all the resources required by an animal or a group for whatever its requirements, or they may be small and established in relation to simply one function: as some fallow deer defend a patch of ground simply as a mating territory to which to attract oestrus females (page 113). Fallow deer are not otherwise territorial in any other aspect of their lives, and even this mating territory is held for only a matter of a few days or weeks.

But a territory may be distinguished from a simple home range – an area in which an animal chooses to live – by virtue of the fact that it is defended against other individuals. In fact, Robert Hinde noted some years ago that the behaviours involved in the establishment and maintenance of a territory could be broadly divided into three: the restriction of all or some types of behaviour to a more

Speckled wood butterflies in spiralling flight as the territory holder defends a sunspot against an intruder

or less clearly-defined area; self-advertisement within that area to announce ownership; defence of that area against some or all other conspecifics.

Establishment of a territory may have a number of possible functions. Familiarity with a site, resulting from site attachment and defence, may assist with exploitation of the area's food resources (through familiarity with their nature and distribution) and with escape from predators simply in the course of day to day existence. Ownership of a territory may be necessary to attract a breeding partner and may facilitate formation and maintenance of a pair bond. It may further assist with defence of the nest site from predators and may also assist in that it reduces possible interference in reproductive activities by other members of your own species. Once the territory is won, you may get on with the business of breeding and rearing young without continued harassment and continued competition.

Indeed, in general terms, territoriality serves to structure competition to a degree, by substituting initial competition for that patch of space for direct competition over individual resources, be they food, mates or nest sites. Indeed the expression of this device of territoriality may actually reduce overall the number of individual competitive interactions each individual must engage in because, once the territory is established, the owner subsequently has relatively undisputed and uninterrupted access to the resources it contains and does not need to continue to compete over each and every individual resource.

Individual distance

Even amongst animals which are not formally territorial, it is notable that each maintains around itself a 'halo' of clear space – and that it reacts aggressively to any other individual who crosses the invisible barrier and trespasses on that free space. All animals defend this free space, this 'personal territory'; we humans do. You only have to watch a video or television footage which films from above a group of people moving, perhaps, along a crowded street. Despite the crowds, despite the fact that people keep changing direction and move around each other, you will notice a clear halo surrounding every individual in that crowd, a clear space that moves with them as they move – rather like the halo associated with the advertisements for a particular brand of instant breakfast cereal. Many of the movements of others, indeed, are to avoid intruding on that personal space – and if anyone inadvertently crosses over it, we ourselves feel threatened, may react aggressively. How after all, do you feel, when a stranger brushes against you or barges into you in the street?

As we have already noted, this expression of individual distance is most apparent in species of animals which regularly or occasionally cluster together into loose aggregations, into herds or flocks: temporary assemblages which are not constant in composition or with permanent membership, but simply expedient at the time, for exploiting particular distributions of food or in defence against predators (Chapter 10). The function of the response is in essence very similar to that of actual territoriality. It preserves around the individual a space which may help prevent direct interference in its activities by other members of the herd or flock, preserving its own area of operation to allow it to feed, or carry out some other activity, uncontested. The reaction arises from a more general reluctance of any animal to be too close to other individuals. As well as reducing to some degree the level of direct competition, it also reduces the risk of them damaging you or, perhaps, because their very closeness crowds you and prevents you from noticing or escaping from a possible predator. It is exactly this same reaction that must be overcome in allowing proximity of a breeding pair for actual copulation – and a whole complexity of courtship signalling has developed to overcome that very natural reluctance to let any other individual within that vital personal space.

Dominance hierarchies

Antagonistic interactions between individual members of more persistent social groupings may become stylised in the development of simple dominance hierarchies, such that through a series of pairwise interactions each individual learns that it may outcompete individual X, but is itself inferior to individual Y. Through a number of such initial trials of strength, the different individuals within a flock or other social group may develop a simple linear hierarchy or rank order of 'status' (or competitive ability), which is learnt. After the initial squabbles implicit in establishing who stands where in the pecking order, subsequent interactions may be reduced to a minimum since, knowing the likely outcome if pushed to a fight, each individual 'gives way' without contest if subsequently challenged by an individual of higher rank.

Displays of dominance in wolves confirm a hierarchy of status; the subordinate individual defuses further potential aggression by a clear display of submissiveness

In practice, such hierarchies are often far from linear and may indeed prove extremely complex, such that A outranks B who outranks C who in turn outranks D, but in fact D may itself outrank B. In addition, the hierarchies established are not definitive in terms of absolute social rank; dominance may change with context, so that one particular ranking of individuals applies in contexts of competition for food, while quite another rank order applies to contests over mating opportunities or some other resource.

It might appear that such hierarchies strongly favour those at the top, while those at the bottom in general lose out, but in practice even the most subordinate individuals may benefit, since aggressive interactions within the group as a whole are much reduced after the initial establishment of the hierarchy, and by giving way without argument when challenged, a subordinate may stand a better chance of escaping without injury and being able to move on to a different patch of resource and resume its previous activity more quickly than if it continually argues the toss. Indeed it can be shown that, for example, food intake even of subordinate individuals within a flock of feeding birds is higher than that of individuals within a similar flock where dominance relationships are unclear. Aggressive interactions still occur, of course; for ranks are not fixed in perpetuity. Individuals of lower rank may repeatedly challenge those above them to win an increased status; mature animals age and become less competitive, while lower-ranking juveniles gain in strength and experience as they themselves mature. The society is thus not without its bickering and its occasional testing and re-testing of relative rank, but in general, aggressive interactions are once again far lower than they would be if each individual had to contest access to every single resource on every single occasion.

Reducing the risks

And indeed this may be a primary function both of territoriality and the development of dominance rank relationships in permanent social groups: that such devices reduce overall the incidence of aggressive interaction. For fighting is costly, in terms of expenditure of both time and energy which could be devoted elsewhere, and in terms of exposing yourself to a real risk of injury, which even if it may not result in death, could cause a measurable loss in fitness and perhaps a reduction in future reproductive success. It is a common observation in nature that aggressive encounters between two individuals of the same species rarely result in fights to the death or, indeed, in serious injury. Often such aggressive encounters seem to have been reduced to formal tournaments or trials of strength, played according to some set of rules. Contests of this type are commonly ritualised in style, involving inefficient weapons or posturing, in place of actual attack. For example, in many snake species, males fight each other by wrestling without using their potentially lethal poison fangs; sheep, goats and antelope contest amongst themselves by crashing together their horns, while they refrain from attacking when an opponent turns away, when they could rush in and spear it in its unprotected flank. In many species of deer, males fight, when they fight at all, by interlocking antlers and wrestling with each other; again, they rarely use the sharp points of their antlers to jab into an opponent's flank.

The risk of injury and loss of fitness through fighting is so strong an evolutionary pressure that animal conflicts have evolved over the generations to reduce the need for costly fights. The actual incidence of fighting can in effect be reduced in two main ways: by reducing the frequency of aggressive encounters, and by reducing the intensity of those encounters that do occur. In the first place, as we have already noted, many animal species reduce the frequency of aggressive interaction by substitution of some 'conventional' goal of rank or status in place of contest over actual material resources. Instead of coming into conflict every time they compete over each and every resource, they compete instead for territory or dominance rank, which in turn then confers automatic right of access to all necessary material resources without further competition. Individuals compete vigorously for territory or rank, but thereafter the actual frequency of aggressive interactions is very low – and far lower than it would be if each pair of individuals fought over every food item, over every mating opportunity.

Of course, not all species have adopted such a device; in many species, individuals do continue to compete over each and every resource. And even in those cases where ownership of territory or rank has been substituted as a conventional goal to 'structure' competition, there must still be some initial competition to establish status in the first place, or whenever ranks are challenged by newcomers. Yet we still rarely see animals using all the various formidable weapons with which they may be equipped against others of the same species. Rather, as we have remarked, we tend to see the development of ritualised tournaments of some description. The intensity of aggression, as well as its frequency, is reduced by engaging in highly-conventionalised trials of strength or show-off displays. There is no need here to resort to group-selectionist arguments to justify this on the grounds that 'serious fighting wouldn't be good for the species'. Rather, one can rationalise the evolution of such conventionalised combat simply in terms of the interests of each individual: there is simply no point in risking injury yourself by engaging in serious fighting if you can resolve the dispute in some other way. He who fights and runs away, lives to fight another day. And even the temptation to cheat and run in with your rapier-

like horns when your opponent turns sideways to you isn't on average a wise manoeuvre, because if everybody were prepared to cheat on the established rules of combat, somebody one day might do the same to you. Simply one must do as you would be done by: as we noted on page 83, the strategy which proves optimal to the individual in the long run depends heavily on what other individuals around it are likely to do in response, and evolution favours those strategies which are optimally advantageous to every individual within a population where that strategy is adopted as a majority strategy (i.e. of all the strategies available, it's the one which will work best, assuming it is adopted by the majority of individuals within the population (and thus is the one you are most likely to encounter in return; see again pages 82–83).

Assessment Strategy

So how may animals resolve disputes by simple display? They may do so if they can determine from that display the probable outcome of any subsequent fight. They must be able to assess from the display or conventionalised tournament what would actually be likely to happen if they escalated that tournament to a full-blown fight.

Any contest between two individuals may be asymmetric in a number of ways. Firstly, and most obviously, there may be an asymmetry in the fighting ability, or what is more formally called the resource holding potential (or RHP) of the two contestants. But there is also a possible asymmetry in the value of the contested resource from the perspective of each contestant: one individual may have more to gain by winning or more to lose in defeat. This, in turn, may be translated into a greater willingness to risk more in contest for that resource. Geoff Parker (the same who studied so intently the mating habits of dung flies, page 80–81) offers a simple explanation for the ritualisation of conflict based on the assessment by each contestant of the likely gains to itself of any particular encounter if it wins, or losses from that encounter if it loses (the payoff) and the probability of winning or losing based on its assessment of the fighting ability of the other contestant. Following such an assessment, the individual most likely to lose and with least to gain from the encounter will withdraw without contest. When I say that each contestant 'assesses' the losses and gains, the value of the resource and the fighting ability of its opponent relative to itself, I do not mean to suggest that it does complicated calculations in its head in some complex cognitive exercise. Rather, as we have discussed elsewhere in this book, evolution has provided it with a simple series of decision rules to obey, which over evolutionary time have proved to be the ones which work best on the average.

Let us consider an example. Let us assume that we are faced with a contest between two individuals for a patch of food of equal value to each. If one individual can demonstrate through display that it is clearly stronger than the other and would win any actual fight over that food patch, the weaker should withdraw without contest. But if the stronger individual is also relatively well-fed at that point in time, while the weaker has not fed for some time, that same food patch may mean much more to the weaker individual than it does to the one of higher potential fighting ability. In which case, in practice, if it came to blows, the less strong individual might nonetheless be prepared to fight harder than the one of higher intrinsic fighting ability, because winning that patch of food means more to it than it does to its opponent. If that 'willingness to fight' (resulting from the fact that the resource in question is of higher value to one individual than another) can also be demonstrated in display, then once again, the contest may be resolved without further escalation.

In general, once an animal has established a territory, it is hard to displace. If another individual challenges the territorial owner for that territory, the territory-holder almost always wins, even if there is some disparity in relative fighting ability. This is because the territory holder has already, by definition, invested heavily in establishing that territory in the first place – and thus has a good deal of past investment to lose (or, if it does lose, has to reinvest all that again in establishing a new territory). The territory is also intrinsically of higher value to the occupant, because he already knows his way around it, has established familiarity with its resources, the distribution of food, cover and so on. To the attacker it simply represents a potential territory with no added value. In consequence, the holder's loss in fitness would exceed the attacker's potential gain. Once again, if this may be revealed in display, the attacker should withdraw, and indeed in general it is true that the territory-holder usually wins.

In fact this is not always the case and one can, by experiment, stack the odds a bit. The essential point in such experimentation, however, is that the outcome is still completely predictable. Adult male house mice (and the related laboratory mouse) establish individual territories whose boundaries are clearly defined with scent marks, from urine and from glands between the toes, which characterise each individual. If two such mice are allowed in the laboratory to establish adjacent territories in isolation and then a door in the partition which physically separates the two territories is removed, the two neighbours have access to each other's space and may explore it. The outcome of any subsequent challenge for ownership is entirely predictable.

If two mice, of equal size and thus equal RHP are allowed to establish themselves in separate territories of equal size and, when the door between them is opened, each has an opportunity to examine and explore the other's territory, we may predict that while there may be mild display between the two males when they meet each other, the outcome will always be the same, and the individual who is at that point in time intruding into the other male's territory will withdraw. Thus if mouse A wanders into the territory of mouse B and encounters mouse B within that territory, mouse A will, in the majority of cases, withdraw to its own territory; by converse if the two were to meet subsequently in mouse A's territory, after further posturing, it will be mouse B who will now withdraw. Both mice have presumably equal fighting ability. Each has a territory of the same size: each has invested in establishment and marking of that territory. Because the territories are of equal size, each therefore stands to gain an equal amount if it were to annex the neighbouring territory; but because each has already invested in the establishment of its own territory, each, on its own patch, stands to lose more. The inequality therefore consists in having more to lose if you surrender your own territory and this asymmetry ensures the territory holder would, if it came to it, be more willing to fight than would the intruder. Both animals can assess this difference and thus the intruder withdraws. Indeed, in a series of real experiments of exactly this form, the holder of the territory in which the encounter occurred immediately dominated the 'intruder' into his territory in 28 of 30 trials.

Aggressive and submissive postures in male mice

Let us ring the changes and establish as isolated neighbours two male mice of markedly different body size. Once the partition door between the territories is opened, everything remains the same as in the previous case: territories are of equal size, each stands to gain an equal amount by annexing the territory of its neighbour; each again has more to lose when contesting ownership of its own territory. But we have now introduced a difference in fighting ability, since in mice, fighting ability is closely correlated with body size. This asymmetry in body size and resultant RHP is such that we may predict that if the mice meet within the territory of the larger male, the smaller individual will at once withdraw. However, if they meet within the territory of the smaller mouse, given the disparity in RHP, we would predict the larger mouse will win any resulting contest, although the dispute may go on appreciably longer in this instance. In actual experiments, however, the outcome was not related to the weight of the competing mice and even in contests between animals of markedly different body mass, territory holders still tended to win any contest.

What might we predict for mice of equal size and thus RHP, but with territories which are appreciably different in size or quality? – perhaps with territory A double the size of territory B? The mice are of equal size, and despite the differences in territory size, each actually stands to lose a similar amount in that if it loses, it loses its entire territory, whatever size it may have been, and be left with nothing. There is, however, a difference in potential gain: Mouse A stands to gain only an additional small area of territory if it were to contest this with mouse B, while B stands to gain a significant extension. Now in territory B, gains to mouse A are less than losses would be to B, while in (larger) territory A, gains to B are significant. We might predict that, on balance, the territory holder would still win the majority of disputes, but any 'takeovers' would be most likely within the larger territory A.

The table summarises the results of a number of such actual encounters, with the odds 'manipulated' just as described here. Contests included in the table are restricted to those where both mice had had an opportunity to explore both territories before their encounter. All disputes were resolved by simple display and subsequent retreat of the loser (or his removal by the experimenter if he had forfeited his own territory). By definition, the majority of encounters will occur in the larger territory, because it is double the available area. From the results it is clear that the territory holder continues to win in the majority of cases; it is notable, however, that the aggressiveness of that territory holder (measured in terms of the intensity of aggression and how quickly it responded to the presence of the 'intruder' by aggressive display) was greater for mice 'owning' a larger territory; correspondingly, the tendency of an intruder to be submissive decreased as the size (value) of the contested territory increased.

Mouse with larger territory wins as owner in that larger territory	Mouse with larger territory wins as intruder into adjacent smaller territory
19	2
Mouse with smaller territory wins as intruder into adjacent larger territory	Mouse with smaller territory wins as owner in that smaller territory
1	8

These analyses suggest that contests where a degree of asymmetry is apparent in either resource-holding potential or pay-off are settled without recourse to actual fights. However, escalated contests do sometimes occur and serious injury may occasionally be suffered by one or both contestants, even when such contests are normally settled in that species by display. We can readily see why this might happen. Escalated contests might arise where the pay-off for winning is very large compared to the risk of injury in fighting or the potential loss due to injury (as for example, as in the illustration above, if the loser of a territorial contest had little chance of finding another territory). Escalation to out-and-out fighting might also arise where there is no obvious asymmetry apparent from display, which might be used to settle the contest in advance; or, finally, where one or other contestant has incomplete or inaccurate information about any such asymmetry in fighting ability or value of the contested resource.

Escalated contests did occasionally arise amongst our laboratory mice contesting adjacent territories but, significantly, only occurred in two situations: either if mice who had lost the 'battle by display' were not removed at once from the forfeited territory, or if the mice encountered each other before each had had the opportunity to explore the other one's territory (and thus had incomplete information about the potential pay-offs). Lack of information might also arise from an inability properly to assess the true fighting ability of the other male. These 'decision rules' are based, as we have noted, on 'average situations' over evolutionary time and while they allow each contestant to assess the probable fighting ability of any opponent based on its body size or other characteristics, it is in practice only assessing what would be the average fighting ability of an animal of that size in that situation, not necessarily the specific fighting ability and commitment of that particular individual. In the light of this imperfect information – or indeed, what appears to be perfect symmetry – the contest may escalate, at least to a stage where the contestants actually have a chance to assess each other's real fighting ability.

Escalation in these last two situations will only occur if the pay-offs and RHP of the two combatants either appear to be equal or cannot be properly assessed. In such cases there is usually only a gradual escalation of conflict, rather than immediate recourse to all-out war; such step-by-step increase in the intensity of battle offers additional information at least about relative fighting abilities and some form of 'limited war' may reveal an asymmetry after all, not apparent from initial peaceable posturing. In that case, once again, the contest can be aborted and defeat accepted before fighting has escalated to the point at which there is yet risk of any serious injury. If, after this limited escalation, asymmetry in RHP or pay-off is still not apparent, then, but only then will the contestants escalate further to an all-out fight; even then, they will only do so if the likely reward to themselves, by winning, significantly outweighs the cost of the fight.

Most contests which are in some way asymmetric, therefore may be resolved by simple, ritualised displays, permitting assessment of the value to each individual of the contested resource and the likelihood of each being able to defend it. In contests in which there is no clear asymmetry by which it may be simply adjudged, or in which information is incomplete or unclear, the conflict may still be resolved without all-out fighting by a gradual escalation to a formal style of 'limited war'. Contests will only escalate to full-scale fighting if no asymmetries become apparent at any stage during the dispute or if the resource is so very valuable that it must be secured and at whatever cost (i.e. that even the potential costs of injury must be accepted).

Assessment plus memory = dominance hierarchy?

These arguments, developed in the main by John Maynard-Smith and Geoff Parker, do seem entirely cogent and do seem to fit very accurately what we may observe of the behaviour of animals in dispute over some contested resource. As long as we may postulate some ability to assess from display any clear discrepancy in fighting ability or likely gains from each encounter, then actual fighting may be avoided – and it would indeed appear that displays are commonly structured to provide exactly that necessary information. Animals may try to 'cheat', of course, puffing themselves up by raising fur or feathers to try to look as large and impressive as possible; but if everybody plays the same game, the underlying discrepancies are still apparent.

And thus any animal meeting another for the first time in competition for some resource may have indeed a realistic chance of assessing its opponent's relative fighting ability before embarking on a fight. But what if we add an element of memory to this? Because most animals live their lives within a relatively limited home range, they are likely to encounter the same individuals over and over again in competition for resources, and this 'limited list of likely opponents' will be still further exaggerated in socially-living animals, where most contests will be amongst the same few members of a small social group. Steve Wickens points out that if encounters are frequent, and if each individual can remember the outcome of all recent contests it has experienced with every other individual with which it does come into contact, then assessment on each subsequent occasion becomes unnecessary, because the relative fighting abilities of contestants are already known from experience, and contests may be resolved even without display. In effect, assessment plus memory, within a limited group of individuals, delivers all the attributes of a permanent dominance hierarchy.

15

How animals navigate

Perhaps one of the most complex problems of animal behaviour is how animals navigate. How do migrant birds travel literally thousands of miles each year to arrive back in the same small patch, to build a mud nest under the same eaves. How do fish like salmon, spawned in fresh water, travel away to mature at sea and then return, many years later to exactly the same watercourse to breed in their turn? How do fragile insects like Monarch butterflies manage to find their way to a very specific wintering site many hundreds of miles away, even though they have never been there before?

At the simplest level, and as we explored in Chapter 2, there are orientation mechanisms which operate as a direct response to some stimulus, reacting by changing the speed of movement, changing the angle of progress, changing the rate of turning, so that the animal's position is altered in relation to the source of the stimulus. We discussed examples of these when talking about behavioural reflexes, explaining for example how the human body louse ends up at the source of some chemical signal in its environment, or why woodlice tend to aggregate in warm moist areas, simply because the speed of movement slows under moist conditions and speeds up in dry or colder conditions. But what we are interested in here is how animals can move long distances to end up in a given location: accurate navigation from place to place.

Over relatively short distances it is clear that most animals are able to orient themselves in relation to the location of memorised landmarks. Homing pigeons, for example, when moved to a new loft have to learn the location of their new home in relation to a new set of such landmarks. When first released in a new area, realising that it is new and that the familiar landmarks of home are missing, they fly around the new loft in ever-increasing circles, (or at ever increasing height) while they learn the local landscape and the position of 'home' in relation to characteristic landmarks. Domestic dogs and cats do much the same in exploratory forays from a new home, although here, olfactory as well as visual landmarks may be important, and here we should note that orientation in this way by memorised landmarks is only something they do over short distances from home. If translocated beyond their area of familiarity, where there are no landmarks that they recognise because they have never been there previously to learn those way-signs, they must use other methods.

This form of landmark navigation is not restricted to more complex organisms such as birds and higher mammals. We have already met the European beewolf (page 11–12) who identifies the location of its nest burrow in just the same way by memorising its position in relation to obvious features of the local landscape. Honeybees, too, orient themselves, in part, in relation to this spatial memory of distinctive landmarks, at least for relatively local journeys. If the bees are trained to an artificial food source which is positioned so that their flight path to and from the hive takes them along the edge of a wood, for example, they rely on this clear landmark rather than using more complex navigational cues. This can be simply and elegantly demonstrated. If the bees are 'trained' to a feeding table so that the flight path from the hive takes them along the edge of a wood running, let us say, from north to south, and then at night, the hive is closed and moved to a new location which has a wood whose edge runs, let us say, from west to east, then on release, the bees will fly along the edge of the wood in search of food, rather than following the previous compass direction.

However, we have also already discussed that honeybees may communicate to others in the hive the position of newly discovered sources of food, through a communication dance – and that the direction of the food source is communicated in relation to the angle of flight which should be taken in relation to the position of the sun. So it is clear that bees can also orient themselves by using the position of the sun itself as some sort of landmark. And this isn't just a short-term use of its current immediate position. The sun, of course, changes its position in the sky through the course of the day, and it turns out that the bees (and other insects, like Monarch butterflies) are perfectly able to compensate for the sun's movement and can thus, in effect, use the sun to navigate in a fixed compass direction. And it is this ability to use the sun (or, for many night-active species of animals, the stars) as a kind of celestial compass that underlies the mechanism of much long-distance, point-to-point navigation in animals.

The experimental evidence for this is compelling. Imagine an experimental set up in which an artificial food source is placed some 150 m due north of a hive of bees and the bees are allowed to become accustomed to coming there for food. If the feeding table (with some bees still on it) is then moved some distance sideways, the bees, when leaving the table, will still fly 150 m due south, even though this no longer gets them home. (In fact, when they realise that they have been 'misled', they then correct, using their knowledge of local landmarks as before). Bees trained to seek food northwards in one place and then moved (at night, in the closed hive) to a place where the sun at noon is as far north of its zenith as it was south of its zenith in the first location, will now, when the hive is reopened, fly south to search for food, showing that it is indeed the sun (and not, for example, the polarisation pattern of light) that they are using as a compass.

Our imaginary hive of bees is now moved to a new location and the entrance to the hive is opened only in the afternoon, for the bees to fly to a feeding table due south. Then the hive is sealed and moved again, but this time the entrance is opened at 7 a.m. The bees still fly south. The fact that they fly south in this new, unknown location shows that they are using a true compass direction, rather than memorised landmarks; more importantly, the fact that they still fly due south shows that they can compensate for the sun's movements during the day, because they have never previously seen the morning sun. They are managing to orient to the selected compass direction (south) in the morning, even though previously they have only had the opportunity to observe and learn the sun's movement during the afternoon, so that they must be able to extrapolate the full movement path of the sun even from observation of only a limited part of that path. Further experiment shows that they can actually extrapolate the complete solar path in any given area from a mere 10 minutes' observation. And this is an animal with a brain the size of a bee's! Or a Monarch butterfly, whose brain, I read, weighs less than 0.02 g.

Birds use celestial compasses too, but in an even more sophisticated way. Many bird species migrate every year over extremely long distances across unfamiliar territory, between widely separated winter and summer ranges. More curious still, young birds hatched in breeding grounds in the north find their way south to exactly the right winter quarters without help from their parents. While in some species, like migrating geese, family parties travel together and thus the young might indeed have learnt the route, in other species the young often leave before the parents, undertaking a journey they have never made before with uncanny accuracy.

It has been known for many years that they use the sun as a compass just as do the bees. The earliest demonstrations were by a scientist Gustav Kramer, who housed migrating European starlings in entirely symmetrical cages with no internal cue to suggest direction. The cages even had feeding hoppers all the way round the outside and perches facing all compass directions so that there were no cues to position within the cages themselves. The birds were found to orientate themselves in a fixed orientation within the cage, as long as they could see the sun, and would take up that same direction whatever the position of the sun in the sky (so that obviously, like the bees, they have some mechanism by which they can compensate for the sun's movement during the day). Kramer showed that if the apparent position of the sun was moved by mirrors, then the orientation of the birds was moved by exactly the same amount.

The birds are clearly orienting themselves by taking up an angle in relation to the sun's position at a certain time of day, as do the bees, and adjusting that angle to the sun as time passes and the sun moves, in order to maintain that constant compass bearing. Such adjustment implies that the birds must have quite a sophisticated internal 'clock' which tells them the time of day, and that they can then adopt the correct angle to the sun's position for that particular hour. Many animals, in fact, have quite accurate internal clocks of this sort, which entrain and control all sorts of behaviour; the clocks are, in fact, not entirely accurate and may gain or lose a little over the course of an entire 24-hour period, but they are 're-set' each day by the times of dawn and dusk.

Kramer showed that if birds clocks were shifted (by keeping them in artificial light and turning the light on at different times to 'start' their day at different times) and these time-shifted birds were then exposed to the real sun, they took up position within the cage at the angle to the sun's actual position which was actually not the correct compass bearing, but which would have been the correct angle from the sun for the time of day they 'believed' it to be.

Birds, of course, commonly continue their migration flight at night as well as during the day; indeed, some species fly only at night. Orientation here appears to be based on the pattern and alignment of the stars rather than their fixed position and interestingly, in this case, experimental birds are not affected by having their internal clocks 'adjusted'. But this is very far from being the end of the story.

In another experiment, the Dutch ornithologist A.C. Perdeck caught a number of wild starlings in the middle of their migration and transported them hundreds of miles south, from the Netherlands to Barcelona in Spain. When released again, the young birds within the flock, inexperienced in the journey, continued their flight in the original compass direction – and actually established completely new wintering grounds in south-eastern France and Spain at about the same distance from their summer home as the usual wintering grounds in Switzerland. But adult, experienced birds were 'aware' of the displacement and, on release, shifted course to a completely new compass direction which returned them accurately to the traditional wintering area. And the young birds, at the end of winter, migrated in a new direction north-eastwards, to take them back in turn to the original breeding grounds where they had been reared.

Since such homing can be accomplished by birds from areas well beyond those of which they have any experience, we can rule out the use of memorised landmarks as cues for direction. And while, up to this point, all we have really been discussing is fixed point orientation – taking up some fixed and predetermined compass direction with respect to the sun or stars and their movement – now we are in the realms of genuine point-to-point navigation. For this to work, the animal concerned must have some conceptual global map and somehow know where it is within that map. And without the benefits of GPS, it must also know the co-ordinates of the place it wishes to get to, to be able to compute an appropriate compass bearing.

But what information is available to a migrating bird – or, come to that, any other animal – to form the co-ordinates of such a grid? Here again, it appears we can develop such a system from accurate perception of the sun's position, an ability to extrapolate its complete daily path from observation of only one small arc of that movement – and from use of an accurate internal clock. The British ornithologist Geoffrey Matthews suggested a mechanism whereby, using only these abilities, a bird might work out where it was at any instant in time in relation to where it wanted to get to (which we'll call 'home'). The concept was inspired – and as far as we know, accurately reflects what the animal actually does.

Matthews pointed out that if a bird is to the south of its home, the position of the sun at its highest point (zenith) will be higher in the sky than it would be at home; to the north of home, the sun at its zenith will appear lower. If a bird can appreciate this difference in height, it will know whether it is to the north or the south of where it wishes to be. It will not, of course, always be taking off when the sun is at its highest, but we have already established that as long as the bird can observe the movement of the sun through at least a small arc – and knows what time of day it is – it can extrapolate, from that small arc of movement, the full solar track and can thus project 'in its mind's eye' the probable position of the sun at the zenith. But the bird also needs to know whether it is to the east or the west of its home. Matthews argued that it can do that too, using its internal clock, by comparing the position of the sun wherever it is, at a given time of day, with how far the sun would be along its track at that same time of day were the bird at home. If the sun is 'ahead' of where it would be at that time of day at home, then the bird must be to the east of its home; if the sun is behind, the bird must be to the west.

Such co-ordination seems amazing, but is in essence extremely simple: all the bird needs is some memory of the sun's track at its home base, an ability to extrapolate the full solar path in any other location in which it finds itself, from observation of a small arc of movement, and an accurate internal clock – all abilities we know that the birds have. The least accurate, in fact, appears to be the internal clock, but it is at least good enough to land the bird in the right general area and, once close enough to home, of course, it can use the memorised landmarks we have talked about before for final adjustments.

Parallel work on night-flying birds suggests that they, too, employ a form of bi-coordinate navigation, using the position of the stars. Working with blackcaps in Germany, the ornithologist Franz Sauer has shown that they do indeed have a sense of position and alter course if they sense that they have been displaced (as might happen naturally, of course, in strong winds). Catching some blackcaps during their autumn migration, Sauer held them in an artificial planetarium. A blackcap at Bremen at 9 p.m. was shown the night sky as it was not due to look until 2 a.m. The bird's internal clock told it that it was definitely 9 p.m., so this sky must obviously be somewhere else. In fact the night sky at 9 p.m. looks like that in Eastern Turkestan, so the bird must assume it has been displaced there and flies due west to get back on course.

All this seems phenomenal, of course, but appears to be true. Or at least it appears that celestial navigation of some form is the main mechanism for true point-to-point navigation. It turns out that it is probably not the only one and that there are 'back-up' systems, but this appears to be the primary method of choice. We know that because migrant birds, whether flying by day or by night, will not fly at all or will become disoriented if skies are cloudy or overcast. Night-flying migrants also tend not to fly if there is a bright moon. But if it is totally dark, they manage fine. It's almost as if they get disoriented if they are still trying to use celestial cues but cannot quite distinguish them; but when using those celestial cues is out of the question, they switch to something else.

Bees in overcast conditions appear still to be able to work out the sun's position by the polarisation plane of the light they can detect and can at least take up a fixed compass direction with respect to the sun's presumed position. Monarch butterflies, however, appear not to be able to use polarisation in this way and whether birds can do so, or whether there is enough information available to provide the two co-ordinates of a bi-coordinate grid, seems uncertain. What birds can do (and what humans can also do, although it takes practice) is use a grid based on the earth's magnetic field. It is not as accurate as using the sun or the night sky, but it serves. While this was suggested some years ago, the idea had been discarded; homing pigeons, flying with small magnets attached to their wings sufficient to disrupt the earth's magnetic field did not seem to be in the slightest bit disoriented and managed fine. In fact, there were two fundamental flaws in such early experiments. Because birds would appear preferentially to use the day or night sky if they can and thus would not be troubling to use magnetic cues on a clear day, disruption of the magnetic field under such conditions would have little effect; secondly, the magnets used were quite strong, sufficient totally to destroy the earth's magnetic field rather than simply to distort it. The experiments were subsequently repeated using weaker magnets which simply distort the earth's field around the bird, such that they are still detectable, but bent askew; fitted with such distorting magnets,

Atlantic salmon navigate successfully back to the rivers and streams of their birth

homing pigeons flown on overcast days with no celestial cues did indeed become disoriented. It appears that magnetism may well provide a back-up system for overcast conditions – a conclusion entirely in keeping with the recognition that many insects, birds and even mammals like ourselves have a small deposit of the material magnetite within the body (in birds and mammals this is located at the base of the beak or nose) which is sensitive to magnetic fields.

Migrating fish also use such navigational systems for long-distance travel; histological and electrophysiological data suggest a magnetite-based mechanism in the nasal cavities of salmonid fish. For them, however, the final stages seem to be achieved by sensitive analysis of the chemical composition of water. Like those of many birds, global navigation systems have only general accuracy rather than absolute precision. But while birds use visual landmarks once they have returned to the correct general area, fish use chemical cues. Once returned to within the general area of their natal stream, homing salmon actually negotiate their way to the correct river mouth and follow the water to the correct tributary of their birth by detecting minute variations in the water quality and chemical constituents of the water and comparing these with an accurate memory of the composition of the water of their natal stream.

16

'If we could talk to the animals...'

My purpose, in this whole exploration of why and how animals behave in the ways that they do, has been to enhance understanding of the form and function of behaviour – so that when you observe animals in your back garden, observe the behaviour of your own pets, or the horses, cattle sheep and pigs in the farmer's field next door - or the behaviour of less familiar creatures in the zoo or on a wildlife documentary, you can better appreciate what they are doing and why.

By necessity (and within the constraints of a brief introduction), we have had space to consider only limited examples – indeed limited themes – within these pages. But in this chapter perhaps, it is time to introduce some other behaviours: to see to what extent we can apply the general principles established so far to interpretation of some other 'observations' we might make – or indeed observations you yourself may have made in the past yet never satisfactorily explained. In this chapter, then, I want to try and apply the same basic principles to some different examples, to help set you on the path of using the knowledge we have shared, in understanding other behaviours you may encounter. To a large extent this will utilise the same principles we have already explored, but in calling on them in new contexts and using them to explain 'new' examples, it may help re-emphasise them and show how they may be applied more generally. In other cases, answers to the questions may introduce some additional ideas, but ideas which are rather specialised to particular situations and thus did not fit comfortably in the thread we have developed to this point.

In this chapter, therefore, I will adopt a slightly different approach to that which we have followed thus far. In this case I propose to rehearse a number of individual questions which have been asked of me over the years by younger, or older, students who have been curious about the things they have observed. The questions which follow are not asked in any particular order.

Why do some animals bury food items and how do they find them again?

A large number of species store (or 'cache') surplus food against a future time of food shortage. This may be a short-term response to an immediate and temporary over-abundance or may be more calculated, in the sense that squirrels or other rodents may cache nuts or grain during autumn in order to have a regular and predictable supply of food over winter; in exactly the same way, beavers store stems and small branches of willow or other deciduous trees in their dams and lodges, as a larder on which they may draw over winter when the surface of their beaver pond may be frozen.

Where the food is stored in a permanent hoard (such as the beaver dam, or stores of grain within the nest burrow of small rodents) it is clearly easy to find when required. But what about those animals which 'scatter-hoard'? Squirrels often bury individual nuts in lots of different places within their range; Eurasian jays, similarly, bury individual acorns. And it would appear that even the simplest animals do have spatial memory (as our beewolf on page 11–12), memorising locations by memorising their relationship to obvious landmarks. How many such individual locations may be stored in the brain is less certain (there must be a limit) and certainly neither squirrels or jays manage to return to excavate all their hoard – a fact that the trees themselves rely on, in that those which are 'forgotten' may then germinate, and the hoarding behaviour – and bad memory – of the hoarder helps them in dispersal of their seeds.

Why do some predators seem to go into a killing 'frenzy' when faced with abundant prey?

We all know the old saying of 'a fox in a hen run'. Surplus killing, as it is more properly known, is not observed in all predators and is in fact most common among the Carnivora and especially in small mustelids (stoats, weasels, pine martens). It may well have a similar function to the reaction to surplus food just described, and certainly surplus-killing foxes usually 'cache' what they cannot eat immediately and return to it at a later time. However, not all the surplus may be cached in this way and it seems probable that the killing 'frenzy' is a result of over-stimulation: faced with an over-abundance of accessible prey, the predator is in effect faced with a 'supernormal stimulus' (page 25), resulting in a behaviour that goes further than simply satisfying the actual needs of the individual.

Why do cats and other animals fluff up their fur when threatened/threatening?

We can offer two complementary explanations for this: one mechanistic and one in terms of adaptiveness, as in the two parts of this book. When threatened, or stressed, the release of the stress-hormone adrenalin causes erection of the hairs in all mammals (and erection of the feathers of birds) as well as other 'involuntary' responses triggered by the sympathetic nervous system, like sweating or preparation of the skeletal muscles for 'fight or flight'. In addition, of course, fluffing up the hair or feathers also makes the animal appear larger to a potential predator or competitor. This may have the effect of deterring the predator from attack and may equally 'persuade' any competitor to back down, because in this last case, assessment of the Resource Holding Potential of a rival (page 142) is indeed largely based on assessment of fighting ability which is itself, in most animals, directly related to body size or body mass.

Can animals lie?

The example above suggests that animals can give 'false' information, at least about their fighting ability. But of course, if everybody plays the same game, everybody fluffs up in the same situation and thus any initial disparity between competitors is preserved. It is, therefore, quite hard to cheat. In truth, most signals offer honest information. The timbre of the rutting call of a mature red deer stag has been shown to be directly related to the size and structure of the pharynx and in consequence gives a very accurate reflection of its body mass – and thus, by extension, its fighting ability. Likewise, we have already noted that the width of the black stripe on the breast of a male great tit is actually a very accurate indicator of its status (page 120). The position of cheats in society is reviewed more formally on pages 134–135 but, in effect, lying can only succeed when liars are relatively uncommon within society.

Which animals use tools?

It used to be considered that tool use was unique to the human animal and defined us as different. More recently, of course, it has been established that many other animals use tools too – and not simply opportunistically, but that they may even craft tools or shape them deliberately for a particular purpose. Chimpanzees have now long been known to use grass stems, or the petioles of leaves, to fish out termites or grubs from termite nests or rotten logs, and may 're-shape' these if they become bent or distorted. Chimps may also chew plant leaves into a pulp to be used as a sponge for absorbing moisture from small cavities so that they may drink. Nor is such behaviour restricted even to our closest cousins among the apes: Galapagos finches also may hold cactus spines in their bills to prise food items out of crevices.

Some tool use may be innate, but the majority seems to be learnt – either anew, through insight learning as discussed on page 53, or by cultural transmission from one individual to another. This last results in marked differences in tool use between different populations even within a given species.

Do animals use medicines?

It would certainly appear that many animals respond to dietary deficiencies or even the accidental ingestion of toxins by seeking appropriate 'medication', making sure that they eat foods rich in whatever nutrient may have been deficient from the diet up to that point, or 'swamping' the toxic effect of some poisonous compound by eating large quantities of other food, to act like blotting paper.

It would appear that in a manner exactly analagous to the simple detection of water deficit (through assessment of the osmotic potential of the blood as it passes through the carotid artery) many animals regularly assess the status and balance of individual nutrients in the diet. If they identify an imbalance or deficiency, they may seek foods which are richer in that missing nutrient. In some cases they can detect that simply by smell or taste; in other cases they must learn what materials contain specific minerals or other nutrients, through trial and error. And for those elements which they may require which are more rare in their environment, animals as distinct as elephants or Scarlet Macaws learn and remember whereabouts within their home range they may find the mineral clays and licks, or the specific herbs which contain those less common elements, and may thus make specific excursions to seek these out. It is indeed surprising just how many different herbivore species actively utilise mineral licks to increase intake of salts which are at low concentration in their usual plant foods or, like the macaws, use clays to neutralise the toxins present in unripe fruit.

Why do birds sing?

I suppose the first thing to note is that not all birds do sing, but of those species which do, or at least make loud calls of some description, it is widely supposed that they do so in order to advertise their ownership of a territory to others, to "stake" possession (and reduce the number of cases in which they may physically have to fight off intruders). In most cases it is primarily the males who sing and many also sing as an advertisement of themselves or their territory, in order to attract a mate.

Indeed, it is notable that birds which sing in the breeding season often do so from a prominent perch from which they will be especially obvious, and that they also sing or call more in the early hours just after sunrise than they do at other times of day. (This same preference for calling in the early morning is also noted for other 'callers', like gibbons or howler monkeys). It appears that the quality of the air at that time is at its best for transmission of sound over long distances; later in the day, as it begins to heat up, the air becomes more turbulent and sounds do not travel so well. Both these observations are entirely consistent with the idea that the calls or songs are for self-advertisement, although amongst monkeys the calls may also function to maintain contact between different members of a dispersed social group.

But many birds continue to sing even later in the season after nestlings have hatched or even fledged. While there may still be a need to advertise and defend territories in order to ensure uncontested access to available food resources and in order to have enough food for the growing youngsters, it has also been suggested that they continue to sing simply

to keep in practice. While it would appear that for most songbirds the basic structure of the song is innate, the detail is learnt and refined through practice, so that it is possible that continuing to sing throughout the season helps to develop and refine individual song patterns. We have already noted that male chaffinches learn at least part of their song by listening to their fathers, so that while there is a basic genetic blueprint for how a chaffinch should sing, details may be refined and learnt and thus there are almost local dialects or variants, as each son learns something idiosyncratic from his father; individual blackbirds' songs change even through a single season, and from one season to the next. But such an argument is a bit 'chicken and egg', for why would the bird need to refine its song? Why does it need to change it from season to season? It could simply be that it continues to sing in territorial defence, or for some other purpose, and that during that repeated singing the song structure changes simply by chance and through over-frequent repetition. But if it does 'rehearse' and modify its song, perhaps sometimes birds do sing merely out of simple pleasure and *joie de vivre*?

Why do some birds mimic the songs of others or even other sounds?

Perhaps the answer to this one is hidden in the question above. Birds of most species inherit a basic song structure and song repertoire, but much of the detail of the song is learnt, either unconsciously, through being exposed to the songs of others, or through progressive practice and refinement. And if the song may be modified by listening to the calls of others (as chaffinches learn from their fathers) perhaps it is not so surprising that some species may even incorporate within the call other sounds they may have heard which are easy to reproduce, like a telephone bell or a police siren.

Do animals need to learn their 'language'?

As we have just discussed, it is clear that for many more complex animals, while the basic structure of the language is innate and genetically pre-programmed, subtlety of expression may indeed be learnt – often producing clear regional 'dialects' in language. But this is only probably true for more complex animals and more complex signals. Where it is important that a signal is immediately understood and its message is relatively simple, or amongst more simple creatures, displays tend to be completely 'pre-set' in the genetic make-up. Indeed, there are advantages that this is the case so that the implication of the signal is clear and incontrovertible and the 'recipient' of such a signal can respond immediately and appropriately. What does have to be learned about 'language', especially in socially living animals, is the correct context of use for different signals. Wolves, for example, have a highly differential repertoire of signals, which often consist of a combination of facial expression, body and vocal signals. Young wolves, of course, have an innate ability to produce a growl – or a bark-sound – but they do learn by experience (feedback from adults) as to when it is appropriate to use such signals.

Why do animals play?

The classic explanation of play behaviour was always that, through play, young animals exercised the muscle systems and particular motor action patterns that they would need as adults. Thus in playing with each other, or pouncing on inanimate objects, predatory carnivores learn and refine the hair-trigger responses they will need in order to capture prey as adults. Domestic cats often bring home live prey (small voles or mice) and release them for kittens to 'play' with. While this behaviour is often distressing to human observers, as the mouse is taunted and caught and re-caught, it is in fact a valid way of ensuring the kittens learn the behaviours required to catch and kill prey for themselves in the future. Social play (rough and tumble play, or mock copulation) amongst a wide variety of mammalian species has also traditionally been explained as a mechanism through which juveniles of more social species may learn the 'correct' social behaviours and social responses which will enable them as adults to integrate and operate within a complex social structure,

European otters playing

learning the correct behaviours for submission or threat, for dominance or appeasement. Indeed, classic behaviourists distinguished clearly between solitary 'object play' which is, on this paradigm at least, simply a matter of motor maturation and 'social play', which may develop both co-ordination and social skills.

But adults also play – or at least appear to. In purely functional explanation, we might simply suggest that such play was a continuation of these same objectives: further practicing and refining motor skills and keeping them honed or, in social play, cementing pair bonds or relationships within more complex social groups. But such an explanation is rather reductionist and increasingly in recent years, some behaviourists have suggested that animals may indeed play simply for the fun of it.

Do animals feel pleasure?

Both in talking about bird song and about play, we have suggested that animals may simply enjoy doing these things. So do animals feel pleasure? Some scientists suggest that animals – at least the more complex ones – must have the capacity to feel pleasure, arguing that pleasure would be adaptive evolutionarily in 'rewarding' animals for making decisions which enhance their fitness, just as pain or suffering 'punish' actions which are not adaptive. Others argue that the successful achievement of some physiological or behavioural goal, following a period of appetitive behaviour towards that objective, may in itself be 'reward' enough, from the point of view of natural selection, without the need to postulate conscious pleasure. This is a difficult question to resolve, but one may perhaps question how the phenomenon of incentive stimulation (page 42) would operate if animals did not derive pleasurable experience from that incentive, or why, in experimental situations, animals will continue to work for 'sweet' rewards, even when they are fed to satiety?

Do animals feel emotions?

As with determining whether or not animals can actually feel pleasure, it is actually very difficult to assess the extent to which they feel emotion. Certainly, on the negative side of things, animals show obvious symptoms of anxiety when placed in unknown and potentially threatening situations, as can be recognised by, for example, pilo-erection (raising of the fur in mammals), increased sweating and more agitated behaviour. It's notable that when animals get used to a given situation, such behaviour gradually disappears. Different individuals vary in the levels of anxiety they may express: in animals, as in humans, we can find high- and low-anxiety traits.

It is now quite widely suggested that animals of a more complex organisation, with distinct cognitive abilities, do experience emotions in much the same way as we do and that these emotional 'feelings' play an important role in the performance of adaptive behaviours which may influence or change that status. As we have suggested above in relation to considerations of 'pleasure', there is a considerable literature to suggest that much of the function of emotion may indeed be to provide a convenient way to reinforce behaviours which are (or were) in some way adaptive, such that performance of these appropriate behaviours is in some sense rewarding. Clearly animals cannot 'know' that performance of a particular behaviour in a given situation is necessarily adaptive in any evolutionary sense, but if performing that behaviour 'feels right' or is in some other way rewarding, then this would act as an evolutionary shortcut, encouraging the animal in the short term to perform behaviours which are indeed, in the longer term, adaptive.

Do animals make friends within a social group?

In trying to be as objective about this as possible, we can certainly establish that within social groups, certain individuals spend significantly more time together with other particular individuals within that group than they do with some others, so that they are not associating with all other individuals within the group at random. But whether this constitutes 'friendship' or may have some purely functional explanation is hard to determine. It could simply result from practical considerations: if you mix more with others close to your own social rank within a dominance hierarchy, for example, or those way above you, there is a greater likelihood of aggressive interaction, whereas if you spend more time with animals close to your rank, but far enough above or below you that the pecking order is clear and obviously recognised by both parties, then the probability of aggressive interaction is decreased.

We do know, however, that not all these associations are necessarily passive and that in some instances animals may form deliberate alliances. During Jane Goodall's long term studies of chimpanzees, one dominant male (Mike) was 'overthrown' by a pair of subordinate animals acting together; neither would have been able to challenge the dominant male alone, but as a coalition they were able to displace him. Similar occurrences have also been reported among lions, where pairs or groups of young males may act together to 'take over' a pride of females. Mature fallow bucks may often tolerate a younger male within their rutting area and in the same way many territorial roe buck have been observed to tolerate so-called' 'satellite' males within their territories, who may even obtain some matings – but the relationship between these is often unknown and it may well be that they are related in some way and thus kin (page 130,131) rather than 'friends'.

Why do some animals have their eyes on the sides of their head and some on the front?

Irrespective of whether we are looking at birds, mammals or some other taxonomic group, it is characteristic that predatory animals tend to have eyes at the front of the head, while prey species have eyes on either side of the head. And this relates to the fact that having the eyes close together at the front of the head means that the visual fields of the two eyes (the field of view) show significant overlap. It is only in such area of overlap of the visual field (or binocular vision) that the animal can effectively judge distance and depth and have the visual acuity necessary to capture live prey. But such binocular vision is at the expense of only being able to see in a comparatively narrow arc in front and means that you cannot see behind you.

By converse, if your eyes are mounted on either side of the head (and each has, say a 180 degree field of vision), you may have only a very small field of binocular overlap, but you can see as well behind you as in front – with such all round vision a tremendous advantage for prey species in detection of potential predators.

Why do bullocks in a herd sometimes mount each other as if in copulation? And why does this sometimes also happen amongst adult cows?

Such behaviours are not restricted to domestic ungulates but may be observed in many wild species as well. Mounting behaviour in fact is not a behaviour restricted to sexual encounters but is quite a common expression of dominance, and in many cases homosexual mounting (same sex mounting) is part of the complex suite of behaviours by which animals in a permanent social group express and maintain dominance over each other. Such behaviours are also sometimes exaggerated if one of the members of a group of adult females has come into season (come into oestrus); in this case, other dominant females in the group may regularly mount that female, although in this case also, the behaviour is primarily one of social dominance.

Why do some animals urinate or defaecate on top of the urine of others?

European badgers defecate in group latrines at territorial boundaries to reinforce links within the group and also to advertise territory boundaries to neighbouring groups

Once again, this may have a dominance or territorial function. Thus the dominant stallion in a group of wild horses will not uncommonly urinate on top of the fresh urine of one of the mares in his harem, or defaecate in close proximity to fresh dung. This is really an advertisement display not so dissimilar to that of birds singing to advertise ownership. In other species (such as muntjac deer, for example) where dung may be used as a marker of territorial boundaries, one individual may deposit dung on top of the dung left as a marker by the holder of the neighbouring territory, as a sign that the marker has been noted and to re-establish his own boundary. In this case, a series of successive defaecations accumulate in a regular latrine.

Latrines are also used by some other species, such as hyaenas or European badgers, who defend a shared territory as a social group; in this case while the group defaecations still may serve as a declaration of the group territory, defaecating in the same latrine well within the territory boundary also exposes each individual to the scent of all the other members of its own social group and 'reminds' it of who are the others of its own social grouping.

Why do some animals wrinkle up their noses after sniffing at the urine or directly at the vulva of other individuals?

Such behaviour is called 'flehmen' and is displayed by a range of animals from primates through to various species of ungulates. It is shown primarily by males but also by dominant females (group leaders) in more social species. Wrinkling the lip in flehmen actually has a physiological function, in that it exposes the openings, tucked just behind the upper lip, of the vomeronasal organ which is an extremely sensitive olfactory organ – capable of detecting very small traces of chemical scents: perhaps only a few molecules. Females in oestrus pass out small quantities of oestrogen and other hormones in the urine and males and other dominant females within the social group smell the urine, and show flehmen, to assess the oestrus state of other females in the group and to assess whether or not they are likely to be receptive to mating.

How much is scent used in animal communication, by comparison to sound or visual signals?

In addition to using the presence of a few molecules of scent to detect the oestrus status of others in the group, males of many species of mammals also use scent, themselves, in courtship or territorial display. We have already noted that badgers and other social carnivores – as well as muntjac deer – may also use latrines to mark territories; many animals also use urine as a territorial marker. But there are many other forms of scent marking or signalling by scent.

Deer, for example, have up to seven or eight different scent-secreting glands distributed over the body. The secretions from these various glands may again be used, by more solitary species, in territory marking: the dik-dik of Africa, or different species of duiker, smear secretions from a gland just below the eye onto twigs and other vegetation as a powerful territory marker. In a similar function, roe deer and other species rub secretions from glands between the antlers onto posts or the boles of trees, or scrape the ground with their forefeet to leave traces of scent from glands situated between the cleaves of the foot. But it is not all to do with advertising territory. Scents may be used for personal advertisement as well: mature males in many mammal species give off a distinctive and very powerful odour during the breeding period which can be detected at some considerable distance. Rutting male red deer or fallow bucks, for example, give off a very powerful odour indeed to advertise their presence to passing females, sometimes even enhancing this by urinating directly onto their bodies, or urinating into mud patches in which they then wallow to plaster themselves with aromatic mud!

But scent is not just used for territorial marking or sex. We have already suggested that sharing a latrine facilitates communication between different members of the same social group of badgers and actual body scent may also be used in the same way by different species

– for recognition of other members of the same social group and distinction of strangers. It has been shown for example that amongst different species of deer, the scent secretions of different glands are not only as individually recognisable as a fingerprint, but also contain coded information on age, sex and social group. The smells released when the hairs of the metatarsal gland (a gland on the back leg) are flared, serve to keep young fawns following the correct mother, while the gland below the eye in young deer calves opens when they are suckling but closes when they are satiated, presumably communicating this quite subtle information to the mother. In a study of communication in the European fallow deer, Ruth Lawson showed that 79% of all interactions between adults involved scent and only 22% used visual cues; by contrast, the majority of interactions between mothers and fawns involved sight (97%), although 22% also involved vocalisations and 18% used scent cues.

Odour is also important in navigation, as we discussed in Chapter 15, in that migrating birds use odour cues as part of their recognition system for home and fish returning to their 'home stream' also use the characteristic chemical 'signature' of the watercourse to find their way to the correct stream in which they were themselves spawned.

What are pheromones?

Pheromones are a special category of chemicals used in communication, in that they do not simply convey information, like other chemical signals, but actually have a direct physiological effect on the recipient. In a sense their action may be considered analogous to that of the action of hormones within an individual's body (in triggering or directing a physiological response) with the difference that these are 'exo-hormones', released by one animal yet having a direct physiological effect on another individual.

Perhaps pheromones are most regularly encountered amongst moths and other insects where females release pheromones which attract males towards them for breeding; many social insects like ants or bees release 'alarm pheromones' when attacked or in danger of attack, which trigger an aggressive response in other members of a colony. But some mammalian hormones and scent signals also act as pheromones. Indeed the presence of a mature male may act to advance oestrus in a number of ungulate species. While in seasonally-breeding species such as red deer, the onset of oestrus is largely entrained by physical condition and changing daylength, the presence of a mature male seems to serve to trigger the onset of oestrus and may even advance it by a considerable number of days.

In the same way, the oestrogen released by a cycling female also may affect the oestrus period of other females around her, such that over a period the oestrus cycles of all females in a social group may be synchronised with each other. They start off totally independent of each other but the pheromones released during each female's oestrus influence the oestrus period of others within the group until they are all closely co-ordinated. This has a real function in wild, free-living animals in that as a result of such oestral synchrony and a common gestation period, all the young in any social group are likely to be born at the same time. This is helpful in the co-ordination of 'nursery care' for those species, such as lions, which may show communal care of the group's cubs, but it also has an anti-predator advantage for prey species – in that, if all the young are born at around the same time, then even though some may be taken by predators, fewer will fall prey overall than if births were not closely synchronised and thus young and vulnerable juveniles were available over a much more protracted period. [7]

But in some species, dominant females can suppress completely the oestrus cycle of subordinates?

Yes, this is also true. Among social insects in particular, the queen actively suppresses the reproductive activity of worker ants or bees and a similar phenomenon is noted even among mammals: in African mole rats (page 101–102) where reproductive activity of worker females is suppressed by pheromonal activity. Reproductive inhibition of subordinates is also noted within many species of social carnivores like hunting dogs or wolves, where only the alpha female in the group produces young. This may also have some pheromonal component but is also due to the fact that the hormonal activity of animals low down in a dominance hierarchy is also suppressed by stress (adrenal hormones have an inhibiting effect on sex hormones); in either case dominant females are able actively to suppress reproductive activity in subordinates. This reduces competition for food for the young and ensures that the alpha female's young have the greatest (uncontested) chance of survival. It also guarantees a good supply of non-reproductive relatives as 'aunts' to share in the care and feeding of the alpha female's young.

---

7 If a given predator can only take one prey item a day and can only catch that prey when it is newly born (if, for example, a jackal can kill only one gazelle fawn each day and can only kill fawns within their first 30 days of life), then if all the births are synchronised to much the same period, it may be limited to 30 fawns in total; if the births were spread out over a longer period, the predator could continue to take one fawn a day for a much longer period, the overall mortality rate would be greatly increased and thus the probability of your own fawn being killed would be correspondingly higher.

Why do many social animals show 'mutual' behaviours in grooming or preening each other?

As with some of the other questions in these pages, we may offer an answer at a host of different levels. In fact, when animals such as monkeys groom the fur of other individuals, they may often be observed to remove loose flakes of skin or salt and ingest them; as well as caring for the fur of the individual being groomed, the groomer may actually be benefiting directly by intake, for example, of salt or other minerals in which the diet may often be deficient (see above). Further, even to the extent that the behaviour does 'care for' the fur of the individual being groomed, by removing foreign objects and possibly parasites, the groomer may still not be being wholly altruistic – in that as we discussed on page 134, there may be a high probability of reciprocation with the groomer becoming the "groomee" in due turn and thus actually obtaining reciprocal benefit. Finally, grooming behaviour is commonly used among social species as one of the suite of behaviours used in establishment and maintenance of dominance hierarchies and social cohesion, so that the behaviour serves to reinforce group membership and reaffirm cohesion of that social group. Mutual grooming (or 'allogrooming') is a fairly common behaviour in many social animals (and not just primates) being observed among horses and in some species of deer, amongst many other species.

Are humans the only species in which sexual intercourse is to a large extent decoupled from procreation?

No. In fact amongst bonobos (a smaller species of chimpanzee relatively recently recognised as distinct from other chimps) soliciting by females and copulation by males is regularly used as a form of social-reinforcing behaviour just as we have described above for mutual grooming. Frequency of copulations rises when an individual returns to a group after an absence, or after an interaction between neighbouring (competing) troops. In a similar way, amongst many nesting birds, especially those which nest colonially (like many sea-birds), there is a great increase in the rate of copulation when one partner returns to the nest after a period of absence (perhaps having been away foraging, or collecting nest material) – to the extent that the pairs copulate at far higher frequency than would be required simply to fertilise any developing eggs.

Classically, this also was seen as part of the bonding process, reinforcing the developing pair bond. However we have discovered (page 125) that cuckoldry is far from uncommon in the animal kingdom, even among supposedly monogamous species. Purely physiologically, the latest ejaculation received displaces much of any sperm which may have been deposited earlier – and thus the urgent and apparently celebratory copulations which accompany the return of an absent partner may in fact have less to do with any such reinforcement of a pair-bond than an urgent need for the male to displace any sperm which may have been deposited in his absence by some extra-pair mating, thus increasing the probability of his paternity of any offspring which may be produced. And people call me cynical?

Do animals still recognise their mother or siblings if they happen to meet again?

When this question is asked it is usually posed in relation to domestic companion animals. In most social species in the wild, females tend to remain with the mother's group after reaching adulthood, so have no opportunity to 'forget' associations. Males may disperse as subadults to join other social groups (usually for mating opportunities), but in this case would rarely return to encounter mother or siblings – unless two brothers leave together and remain together as a coalition (above). In less social species, individuals simply disperse to establish their own home range or territory so, once again, are unlikely to re-encounter their mother or litter-mates.

But in some species, such as chimpanzees above and some social carnivores like wolves and dogs, individuals which have left their natal group may return after some months or years of absence, and it would appear in such cases that they are recognised. Amongst domestic pets, too, (and, again, especially notably among dogs) it is common experience that if reunited with litter-mates or their mother, individuals do appear to recognise each other.

We have already hinted that chemical communication (communication by scent) may play an important role in animal communication and indeed this is especially true in relation to individual recognition. Among vertebrates, at least, we know that every individual has a distinct smell and that mother and offspring use that individual smell to recognise each other during the juvenile phase of dependency. What we have not yet discussed is that some of the chemicals emanating from mother to offspring are actively pheromonal and are known to induce positive emotions ('appeasing pheromones'); further, that there is a direct link between distinct sensory systems and those brain areas regulating emotions and that emotional memory is known to be very strong and persistent. Thus, if an adult happens to meet its mother, it is likely that it will recognise her (positively, since her specific smell will elicit positive emotional memory).

Do animals have personalities?

Again, this question is usually asked in relation to domestic pets, but we can certainly answer it more generally in relation to the extent that individuality exists in any species. (What we may anthropomorphically denote 'personality' in effect simply describes individual variation in the way animals respond to a given stimulus situation). It will be no surprise to learn that the answer is conditional and that there is comparatively little individual variation in behaviour among those animals whose actions are largely controlled by simple reflex or innate action patterns. Individuality only begins to emerge within those species whose brain and central nervous system have reached a level of sophistication which permits more conscious control of behaviour or cognitive decision making.

Observation of individual animals over more protracted periods of time has in recent decades emphasised that degree of variation in response, with examples presented for, for example dogs and cats, hyaenas, bears and great apes such as chimpanzees and orang utans. In some cases, recognition of such differences in 'personality' has been somewhat subjective and qualitative, with a risk perhaps of projection onto the animals observed, of some anthropomorphic judgement of the observer, but there are a number of more formal analyses of this sort of variation in response to a given situation. In fact, in recent years a variety of personality tests have been developed for domestic animals like dogs and horses, but even in, for example, inbred mice it is known that individual behavioural responses to the same test situation show considerable variation.

Are animals capable of self-awareness?

Given such individual variation, are animals able to recognise themselves? In many species yes, it is clear that animals have some sense of self and self-recognition. Indeed one of the simplest mechanisms that territorial species which use scent marks, may use to detect an intruder in their territory is 'scent-matching': comparing the smell of any recent scent-mark they may encounter with their own smell. This ability to distinguish self from non-self is actually the simplest mechanism for the detection of an intruder, without needing any more complex memory of the scents of other individuals within the environment; scent matching and the ability to recognise self are also of enormous value to neighbouring individuals in recognising when they may have strayed from their own territory into that of another.

Visual recognition of self is also reported for a number of species. Not only among primates, but also dolphins and other cetaceans and some birds, it is reported that individuals clearly react differently to an image of themselves in a mirror than to catching sight of another individual – and that that difference in response does relate to recognition that the mirror image corresponds to 'self'.

Does an animal remember its life history and what has happened to it over time?

By very definition, the process of learning is of course a function of memory: remembering how to behave in given circumstances, remembering the likely consequences of a given action. So yes, memory of, at least, the consequences of past events is fundamental to controlling future behaviour. Memorising life events is essential for survival, of course: remembering where food can be found, where danger may occur and modifying behaviour accordingly. In some instances this may involve very long-term feats of memory – when a particular resource is required only very occasionally (as elephants may remember over many years the position of distant waterholes needed only in instances of extreme drought). But how conscious may be that memory, how 'abstract' may be the recollection of its past life is hard to determine. However, given what we have just established above that many animals may have a clear awareness of 'self', it doesn't seem impossible that they might have other 'abstract' concepts as well.

Do animals dream?

Once we start talking about those species in which there is greater and greater cognitive control over behaviour, we may start to wonder how complex that brain function might be and how similar it is to our own. And yes, there is good evidence that many species with

more complex brains do dream (or at least show periods of rapid eye movement while asleep, and the brain waves which correspond to dreaming in humans).

Does this imply that their brains are indeed able to represent memories of earlier events 'virtually'? We do not know whether animals ever 'sit and think about their life'. Some argue that there is no – or hardly any – evidence of consciousness is animals and thus we shouldn't suppose that they are able to 'reflect' in any abstract sense. Others argue that, by converse, we cannot prove that animals don't have consciousness in that sense and thus, given that at least some animals have brains that are quite comparable to our brains, we may presuppose that they may do.

Do animals make plans?

A number of more social species can certainly show co-operation in foraging activity, particularly in hunting. It is clear that African hunting dogs, for example, co-ordinate their activities during a hunt; killer whales co-ordinate actions to topple seals into the water from ice floes; whales and dolphins may corral fish by creating a visual net of air-bubbles; pelicans also work together to surround a shoal so that all may feed. Whether or not this is simply a series of responses to ongoing events or involves forward planning is hard to establish. But the coalitions which may be formed, for example, by pairs of chimpanzees in order to take over a social group from another individual dominant male (page 162) are not something that simply takes place as a spontaneous event, but would appear to be final act of a series of actions over weeks or months. Thus, strategies certainly exist and can be aimed at a distinct goal. Yet again however, whether this is as the result of conscious forward planning, we cannot know?

End piece

And as all these questions stretch further and further into the more philosophical, as it were, it is harder and harder to offer definitive answer, because sadly, unlike Dr Doolittle, we cannot talk to the animals. What we might end with is the recognition that, while for animals with less sophisticated nervous systems, much behaviour is fixed, pre-programmed (innate) and largely inflexible, in those species with more complex nervous systems and especially with complex brains, behaviour becomes less predictable, more variable. With the possibility of the development of novel behaviour through insight learning, with increased individual variability, increased potential for cognitive decision-making in response to analysis of information rather than simply as a reflex response, then clearly the behavioural repertoire of more complex animals goes far beyond instinct. What we do know is that this increased complexity of both brain function and behaviour is closely correlated with social organisation, with the greatest complexity developed among the most social species which need that complexity and subtlety of response for social communication and social co-operation. But we may never know if they truly 'think' like us.

Suggestions for further reading

Dawkins, M.S. 1986 *Unravelling Animal Behaviour,* Longman

Krebs, J.R. and Davies N.B. 1993 *An Introduction to Behavioural Ecology,* 4th edition

Ridley, M. 1995 *Animal Behaviour*. Blackwell Scientific Publications, Oxford

Index of keywords

Index of species mentioned in the text